Lecture Notes in Physics

Lecture Notes in Physics

Edited by J. Ehlers, München, K. Hepp, Zürich
R. Kippenhahn, München, H. A. Weidenmüller, Heidelberg
and J. Zittartz, Köln
Managing Editor: W. Beiglböck, Heidelberg

143

Present Status and Aims of Quantum Electrodynamics

Proceedings of the Symposion
Held at Mainz University
May 9–10, 1980

Edited by G. Gräff, E. Klempt, and G. Werth

Springer-Verlag
Berlin Heidelberg GmbH 1981

Editors

Gernot Gräff
Eberhard Klempt
Günter Werth
Institut für Physik, Johannes Gutenberg-Universität
Jakob-Welder-Weg 11, D-6500 Mainz

ISBN 978-3-540-10847-4

Library of Congress Cataloging in Publication Data Main entry under title:
Present status and aims of quantum electrodynamics. (Lecture notes in physics;
143) Bibliography: p. Includes index. 1. Quantum electrodynamics--Congresses.
I. Gräff, Gernot, 1929-. II. Klempt, E. (Eberhard), 1939-. III. Werth, G. (Günter),
1938-. IV. Series. QC679.P73 537.6 81-9161
ISBN 978-3-540-10847-4   ISBN 978-3-540-38753-4 (eBook)
DOI 10.1007/978-3-540-38753-4

2153/3140-543210

CONTENT

Foreword

Since the first measurements of the Lamb shift and the anomalous magnetic moment of the electron,tests of quantum electrodynamics have become a continuous challenge for many experimental physicists. Several times during these years discrepancies between the predictions of the theory and the experimental data have been published, stimulating intense discussions about the physical grounds and mathematical methods of QED. Further improvements of experimental accuracy combined with more careful analysis of the experiments and calculation of higher-order contributions have led again and again to an agreement between the experimental data and the predictions of QED. From the small discrepancies still present at this time, nobody would deduce a breakdown of QED theory. However, regarding the fundamental importance and the model character of QED, further tests with larger momentum transfers and higher precision that check on the validity of higher-order contributions seem highly desirable. Therefore we felt that the time has come for a discussion of the following topics:

- physical ground and mathematical methods of QED,
- mutual relations between theory and experiment,
- analysis of experimental data as being presented today,
- possible improvements of tests of QED regarding experimental aspects and
- contributions of other interactions.

We were very pleased that so many experts actively engaged in this field supported our suggestion to hold a Symposion on the Present Status and Aims of Quantum Electrodynamics at Mainz.

As far as the theory is concerned, the contributions discuss fundamental problems of QED, aspects of unified field theories, relations between theory and experiment, and examples of numerical calculations of QED interactions at large momentum transfer and corrections of higher order in α and $Z\alpha$. However, the major part of the contributions assesses the QED tests at high energies and represents the current status of precision experiments on bound systems and free particles at low energies. Of course, within the time allotted the total spectrum of QED could not be covered. Several important topics, such as the interactions with real photons or macroscopic QED, had to be omitted.

As a result, the symposion revealed some general trends and problems. At high energies new experiments with an even larger momentum transfer may be realised in the future. In contrast, however, further improvement of accuracy in precision experiments is now often limited by the finite lifetime of the system under investigation (e.g., positronium); also the comparison between experimental data and theory is becoming more difficult due to the not precisely predictable contribution of other interactions (consider, e.g., the anomalous magnetic moment of the muon). Therefore the general trend is characterised by the measurement of the QED properties of stabile systems (electron, positron), by investigations of (n'S - nS) transitions, or by the exploration of systems with high Z values, especially hydrogen-like ions and muonic atoms. The reader will find these trends in several articles.

The editors thank all contributors for their engagement. We gratefully acknowledge the hospitality of the Akademie der Wissenschaften und der Literatur zu Mainz. We are indebted to the Johannes-Gutenberg-Universität Mainz, the Verband der Freunde der Universität Mainz and the Regionalverband Rheinhessen der Deutschen Physikalischen Gesellschaft for their financial support.

Mainz, May 1981   G. Gräff

                                                         E. Klempt

                                                         G. Werth

QUANTUM ELECTRODYNAMICS WITHIN THE FRAMEWORK
OF UNIFIED FIELD THEORIES*

Herbert Pietschmann
Institut für Theoretische Physik
Universität Wien

1.   Historical Background

In 1930, P.A.M. Dirac found the relativistically invariant equation of motion for an electron in an electromagnetic field A_μ

$$(i \gamma^\mu \delta_\mu - e \gamma^\mu A_\mu - m) \psi(x) = 0 \tag{1}$$

or

$$(i \not{\partial} - m) \psi(x) = e \not{A} \psi(x) . \tag{1'}$$

(Eqs. (1) and (1') should also serve to define the notation used subsequently.)

Together with Maxwell's equations

$$\square A_\mu = e j_\mu(x) \tag{2}$$

they form the fundamental set of equations for the theory of photons and electrons. From them, Dirac derived his famous hole theory[1] which he first interpreted as a theory for electrons and protons. After it was shown that this leads to an unacceptable instability of matter, the theory was re-interpreted as one for electrons and positrons. The notion of antiparticles was thus created and the discovery of a positive electron by Anderson[2] made independently of theoretical developments led our understanding of the elements of matter to one of its greatest triumphs.

* Supported in part by "Fonds zur Förderung der wissenschaftlichen Forschung in Österreich", Project Nr. 3800.

In spite of these exciting discoveries in the old days, the birth-day of Quantum Electrodynamics is usually associated with the first successful calculation of a higher order correction. In 1948, J. Schwinger computed the anomalous magnetic moment of the electron[3] to be

$$a_e = \frac{\Delta\mu_e}{\mu_e} = \frac{\alpha}{2\pi} . \tag{3}$$

Today we understand Quantum Electrodynamics to be the theory of charged leptons and the photon. It is defined by the Lagrangian

$$L_{QED} = \sum_1 \{\bar{\psi}_1 (i\not{\partial} - m) \psi_1 + e A^\mu j_\mu^1\} - \frac{1}{4} F_{\mu\nu} F^{\mu\nu} \tag{4}$$

with the electromagnetic current of leptons

$$j_\mu^1 = \bar{\psi}_1 \gamma_\mu \psi_1 . \tag{5}$$

The sum goes over the 3 known charged leptons e, μ and τ.

Table 1 summarizes those static properties of these 3 leptons, which are not equal for all three, namely mass and lifetime.

Table 1: Mass and lifetime of the charged leptons

1	m_1 (MeV)	τ_1 (sec)
e	0.511 003 4 (14)	∞
μ	105.659 46 (24)	$2,197\ 134\ (77)\cdot 10^{-6}$
τ	$1782\ ^{+\ 3}_{-\ 4}$	$< 2.3\cdot 10^{-13}$ (theor: $2.8\cdot 10^{-13}$)

It can be seen from eq. (4) that the mass is indeed the only basic quantity in which the 3 leptons differ. (Since there are no transitions between leptons in eq. (4), lepton number is a good quantum number and we could say that they also differ in this quantity.) There is a uni-versality principle, called "μ-e-τ universality", which is only broken by the difference in masses. The difference in lifetime is a direct consequence of this mass difference. Indeed, the theoretical prediction of the lifetime of the τ in Table 1 is based on e-μ-τ universality.

The point-like nature of lepton-photon interactions as predicted from eq. (4) is today tested to the breathtaking limit of about $4 \cdot 10^{-16}$ cm for all leptons. A special section of this conference will give more information on this point.

2. Limits of Applicability of Quantum Electrodynamics

The different lifetime of leptons is not the only consequence of mass differences. There are more subtle effects also, all of which can be computed from eq. (4). Schwinger's correction to the magnetic moment as given in eq. (3) holds for all three types of leptons because it is the lowest order correction in which no mass ratios enter. But if we go to the next order, differences do show up.

$$a_e = \frac{1}{2} \frac{\alpha}{\pi} - 0.328 \ 48 \ (\frac{\alpha}{\pi})^2 + \ldots \qquad (6e)$$

$$a_\mu = \frac{1}{2} \frac{\alpha}{\pi} + 0.765 \ 78 \ (\frac{\alpha}{\pi})^2 + \ldots \qquad (6\mu)$$

The difference is due to contributions from a class of Feynman-graphs, a typical one being shown in Fig. 1. When the lepton of the close loop differs from the external lepton whose magnetic moment is measured, the mass ratio enters, causing differences in the contribution to a_l for different l.

But this class of graphs also leads us to the first limit of Quantum Electrodynamics. For the closed loop does not have to be a lepton; it can also be a hadron (or a quark). In this case, the contribution can no longer be derived from eq. (4). Thus a natural limit is reached at a precision, in which these hadronic contributions become important. We can then either use measurements from hadron physics to compute the contributions or we can use precision measurement of QED to set limits on hadronic quantities. In either case, a comparison of theory and experiments bears no longer exclusively on QED.

A similar type of graph gives rise to the second limit of QED which we shall discuss presently. The graph is shown in Fig. 2; it stems from weak interactions, in which leptons do participate also. Of course, its contribution is expected to be small, but a computation gives infinite result. As soon as we allow neutrinos into the picture, more difficulties arise. Within QED itself, all infinities can be buried into unobservable quantities such as bare masses or coupling constants. Radiative corrections to the weak coupling constants are - in general - infinite. True that in the purely leptonic case of μ-decay or τ-decay, radiative corrections are finite, but this is rather a coincidence than a deep phenomenon. It is due to the good fortune, that a Fierz transformation allows us to collect the charged particles into one current alone by

$$\bar{\nu}_\mu \, \gamma_\lambda (1 + \gamma_5)\mu \; \bar{e} \, \gamma^\lambda (1 + \gamma_5)\nu_e = \bar{e} \, \gamma_\lambda (1 + \gamma_5)\mu \; \bar{\nu}_\mu \, \gamma^\lambda (1 + \gamma_5)\nu_e \; . \tag{7}$$

(Thus the vertex correction of QED can be applied as the only radiative correction.) As soon as we take into account the finite mass of the intermediate boson or we turn to n-decay, infinities pop up.

Thus we arrive at the second limit of pure QED: its connection to weak interactions, typically demonstrated by the contribution of Fig.2.

3.   Unification of QED with Weak Interactions

The second limit of pure QED has been overcome by the beautiful theory of unified electro-weak interactions of Glashow, Salam, Weinberg (and others). It is a renormalizable theory, so that no infinities occur in measurable quantities (except for electromagnetic mass differences of hadrons). Nothing is changed in QED proper, the Lagrangian (4) remains identically the same. Also, ordinary charged current weak interactions are taken over from the good old V-A theory.

$$L_{CC} = \sum_i \frac{g}{2\sqrt{2}} \, \bar{\Psi}_i \, \gamma_\lambda (1 + \gamma_5)\Psi_{\nu_i} \, W^\lambda + h.c. \tag{8}$$

Due to the marriage with QED, however, g is now related to the electric charge

$$e = g \sin \theta_W \; . \tag{9}$$

θ_W is the weak mixing angle and in writing eq. (9) we have merely replaced g by another parameter, θ_W. But θ_W is - for all practical purposes - the only new free parameter to be determined once and for all by experiment. It will occur over and over again; thus eq. (9) may be taken as its definition so that consequent relations actually reduce the number of free parameters.

A characteristic feature of electro-weak interactions (or Quantum Leptodynamics to be extended to Quantum Flavourdynamics when hadrons are incorporated) is the presence of the weak neutral current. Its prediction and verification was one of the corner stones on which the whole framework rests. The neutral current Lagrangian is

$$L_{NC} = \sum_{l} \frac{g}{4 \cos \theta_W} [\bar{\psi}_{\nu_l} \gamma_\lambda (1 + \gamma_5) \psi_{\nu_l} - \bar{\psi}_l \gamma_\lambda (C_V + \gamma_5) \psi_l] Z^\lambda \qquad (10)$$

with

$$C_V = 1 - 4 \sin^2 \theta_W . \qquad (11)$$

Z is the neutral equivalent of the charged intermediate boson W. From neutral current interactions, the value of the weak mixing angle can be determined and the best world value at present is[4]

$$\sin^2 \theta_W = 0.230 \pm 0.009 . \qquad (12)$$

It is precisely the existence of this additional part of weak interactions, which allows finite predictions. In our example, the infinite contribution to the anomalous magnetic moment from the charged intermediate boson (as shown in Fig. 2), another graph contributes due to eq. (10). It is shown in Fig. 3. The most divergent contributions are equal with opposite signs and thus cancel. To render all predictions of physical processes completely finite (thus to possess a renormalizable theory) needs yet another piece added to the Lagrangian; we will deal with it shortly. But before, let us understand the other aspect of electro-weak theory: the unification of electromagnetic and weak interactions. To see this in physical terms, let us look at the second limit of applicability of pure QED, as we have defined it in section 2. Electro-weak theory extends beyond that limit, containing QED as a special case in much the same way that special relativity extends beyond Newtonian mechanics. The limit is reached, whenever v/c approaches unity. In our case, deviations from pure QED will typically show up, when the ratio of typical energies over the mass of the neutral intermediate boson reaches unity or when the prediction of the experiments reaches distances comparable to the Compton wave length of the Z^o. In spite of the very successful high energy experiments with neutral current neutrino interactions (leading to the result of eq. (12)), I personally think that the most direct way of approaching the limit of QED given by eq. (10) is to find its effects in the atomic shell. Though contributions of eq. (10) at low energies are of course exceedingly small, hope to find them lies in the fact that it contains parity violating parts which can be separated from the main contribution of eq. (4). Indeed, in the static limit for an electron orbiting a nucleus of Z protons and N neutrons, the parity violating potential (neglecting nuclear spin effects) can be obtained from eq. (10) to be

$$V_{p.v.} = \frac{g^2}{32 \, m_W^2 \, m_e} \{(\vec{\sigma}\vec{p})\delta^{(3)}(r) + \delta^{(3)}(r)(\vec{\sigma}\vec{p})\} \, Q_W(Z,N) \tag{13}$$

with

$$Q_W = (1 - 4 \sin^2\theta_W)Z - N . \tag{14}$$

Experiments to find effects of eq. (13) are obviously difficult. Many have been planned[5] and in the case of Thalluim, a positive conclusion has been reached[6]. In the case of Bismuth, the situation is more confusing, because 4 groups are working on the problem and their results are not unanimous. But there is some trend as time passes by and the situation is summarized in Fig. 4 (as laid out at the neutrino conference in Erice 1980 by Barkov).

Let us now turn to the problem of finiteness of predictions (or renormalizability of the theory). Again, we should approach this aspect from a phenomenological point of view. If the Lagrangian of Quantum-Leptodynamics contains mass terms, infinite results will appear. Thus the masses have to be generated by some special procedure, which is called Higgs-mechanism (or spontaneous symmetry breaking). Equivalently, we can say that in order to cancel all infinities, charged leptons and the heavy intermediate bosons have to couple to a neutral scalar field H (the Higgs-field) in precisely such a way, that this cancellation is guaranteed. Of course, the way to obtain just this coupling is by going through all the steps of spontaneous symmetry breaking[7]. We shall not do this here, but just give the result:

$$L_Y = \frac{g^2}{4} (W_\mu^\dagger W^\mu + \frac{1}{2 \cos^2\theta_W} Z_\mu Z^\mu)(H^2 + 2\lambda H) + \sum_1 \frac{m_1}{\lambda} \bar{\Psi}_1 \Psi_1 H \tag{15}$$

where

$$\lambda = \frac{2m_W}{g} . \tag{16}$$

Moreover, both the Higgs field as well as the intermediate bosons have to self-interact in the following way:

$$L_H = - m_H^2 (\frac{1}{2\lambda} H^3 + \frac{1}{8\lambda^2} H^4) \tag{17}$$

and

$$L_{WZ} = g \ (\delta^\mu \ \vec{W}^\nu) \ [\vec{W}_\mu \times \vec{W}_\nu] + \frac{g^2}{4}\{(\vec{W}_\mu \ \vec{W}^\mu)^2 - (\vec{W}_\mu \ \vec{W}_\nu)(\vec{W}^\mu \ \vec{W}^\nu)\} \qquad (18)$$

where the fields $\vec{W}$ are related to the physical bosons A, $W^\pm$, Z by

$$W_1^\mu + i \ W_2^\mu = \sqrt{2} \ W^\mu$$

$$(19)$$

$$W_3^\mu = Z^\mu \cos \theta_W - A^\mu \sin \theta_W \ .$$

It remains to mention the electromagnetic interaction of the $W^\pm$, $L_{W\gamma}$, and the free Lagrangian, L_o, of the Higgs field and the intermediate bosons, including the usual mass terms. We then have the full Lagrangian of Quantum-Leptodynamics as the sum of the terms

$$L_{QLD} = L_o + L_{W\gamma} + L_{QED} + L_{WZ} + L_{CC} + L_{NC} + L_Y + L_H \ . \qquad (20)$$

QED remains unchanged but appears as a well-integrated part of a larger scheme, unified electro-weak interactions. As we can infer from eq. (15), e-μ-τ universality is broken in this wider scheme not only by the mass terms, but also by the coupling to the Higgs-boson. It is therefore a challenging and rewarding task to search for and hopefully to find this exotic particle which is predicted by the wider scheme of electro-weak interactions.

4.   Larger Schemes

So far, we have placed QED proper in the framework of the unified field theory. But QED would not be the successful theory it is if it could not describe electromagnetic interactions of hadrons also. Likewise, Quantum-Leptodynamics can be extended to QFD, Quantum Flavourdynamics, by incorporating quarks. Of course, predictions are no longer as precise and parameter-free as they are for leptons, just as in the case of QED. But we can learn a lot about the structure of hadrons by using electrons and neutrinos as probes.

Here is not the place to dwell upon all details of electro-weak interactions of quarks. Let me just give as example the neutral current interaction of quarks in analogy to eq. (10)

$$L^q_{NC} = \frac{g}{4 \cos \theta_W} \left\{ \sum_q \bar{\psi}_q \gamma_\lambda (C'_V + \gamma_5) \psi_q + \sum_{q'} \bar{\psi}_{q'} \gamma_\lambda (C''_V + \gamma_5) \psi_{q'} \right\} Z^\lambda \qquad (21)$$

where the sum over q goes over the quarks with charge 2/3, i.e. u, c, t (up, charme, top) and the sum over q' goes over the charge - 1/3 quarks d, s, b (down, strange, bottom). Also,

$$C'_V = 1 - \frac{8}{3}\sin^2\theta_W$$

$$(22)$$

$$C''_V = 1 - \frac{4}{3}\sin^2\theta_W \ .$$

A comparison of eqs. (21) and (22) with eqs. (10) and (11) shows quite clearly, that the generalization to quarks is quite straightforward.

The unsolved problem remains, how the quarks combine into hadrons. But the concept of gauge theories has also helped in this respect for we picture strong interactions of hadrons as resulting from Quantum Chromodynamics (QCD), the interaction of colored quarks with gauge gluons. Quarks should come in 3 different colors, interacting in an SU(3) symmetric way with 8 gluons. At low energies, this interaction should lead to a potential which confines quarks and gluons unless they form a singlet configuration, which is a hadron.

To conclude, let me just mention that unification schemes for QFD and QCD are also proposed. They involve even larger groups, the minimal being SU(5). We now enter the realm of speculation, in particular since we started from such a firm ground as QED. But there is a crucial ex-periment to test these schemes: They unanimously predict the decay of the proton. Let us hope, that experiments can settle this question without evasion by theorists who push the predicted lifetime ahead of the empirical limit in a snow-plough like fashion.

The phantastic accuracy which has been reached by QED was certain-ly one of the origins for the assurance, that renormalizable gauge theories have still much more predictive power, which started the cumbersome way leading to the victory of unified field theories. It may be the time to use the skills and the endurance which was needed in QED work also in the wider scheme of electro-weak processes, so that even more powerful physics may emerge in the future.

References

1. P.A.M. Dirac, Proc. Royal Soc. A126, 360 (1930)

2. C.D. Anderson, Science 76, 238 (1932), Phys. Rev. 43, 491 (1933)

3. J. Schwinger, Phys. Rev. 73, 416 (1948)

4. I. Liede and M. Roos, Nucl. Phys. B167, 397 (1980)

5. M.A. Bouchiat and C.C. Bouchiat, Phys. Lett. 48B, 111 (1974)

6. R. Conti, P. Bucksbaum, S. Chu, E. Commins, L. Hunter, Phys. Rev. Lett. 42, 343 (1979)

7. See for example H. Pietschmann, Acta Phys. Austr. Suppl. 19, 5 (1978)

Fig. 1:   Typical Feynman-graph contributing to the difference in anoma-
          lous magnetic moment of e and μ:

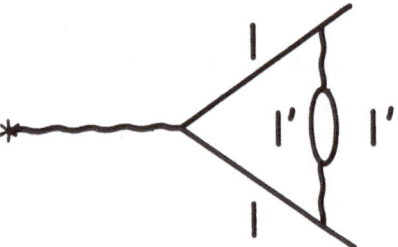

Fig. 2:   Weak contribution to the anomalous magnetic moment of leptons:

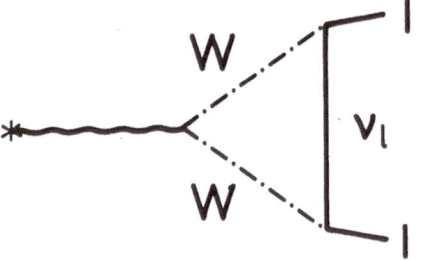

Fig. 3:   Neutral current contribution to the anomalous magnetic moment
          of leptons:

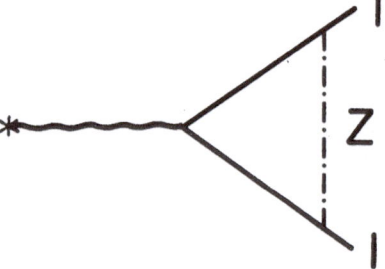

Fig. 4:   Ratio of experimental results to theoretical predictions on Bi
          from 4 groups as function of time:

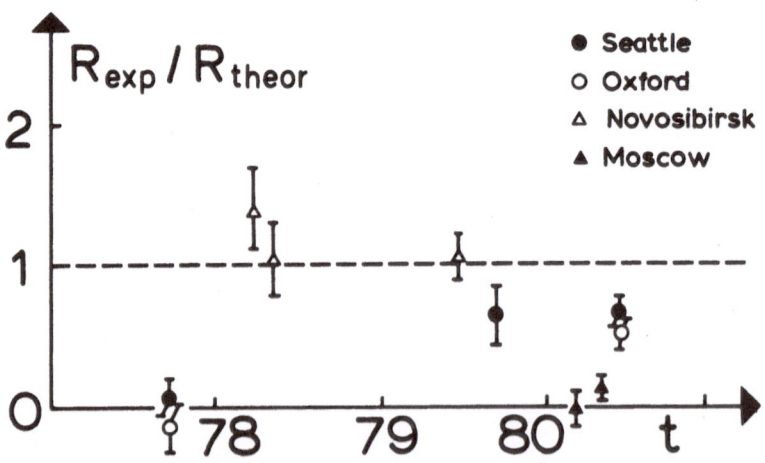

UNIVERSALITY OF LEPTON INTERACTIONS

F. Scheck
Institut für Physik
Johannes Gutenberg Universität, Mainz

1. Introduction

To the best of our knowledge the electromagnetic and weak interactions
of the electron, the muon, and the τ-lepton are universal:
e^-, μ^-, τ^- carry exactly the same electric and weak charges through which
they couple to the photon and the charged and neutral heavy
vector bosons of weak interactions, respectively. Similarly the
neutrinos ν_e, ν_μ, and ν_τ that accompany these charged leptons, seem to
have identical couplings to the weak vector bosons. Within each of
these groups (e, μ, τ), $(\nu_e, \nu_\mu, \nu_\tau)$ all qualitative differences in
decay widths, cross sections, and any other static or dynamic
properties of these particles are due exclusively to the differences in
mass of e, μ, and τ and to the different kinematics following from the
mass differences.

The universality of electric and weak <u>charged-changing</u> interactions of electrons and muons has been known empirically for a long
time and was tested in various experiments[1] (see below). The
properties and interactions of the τ-lepton - even though they are
known much less accurately than those of the electron and the muon -
seem to fit well into this scheme of universal lepton interactions[2].
As yet no conflicting counterevidence has been found.

The principle of universality which also seems to hold for the
interactions of hadrons (although in a generalized frame), is not really understood theoretically. It is not (yet) a necessary and
unavoidable building-principle of a theory of leptons (and hadrons).
However, we do have at least a hint at universality from gauge
theories: Non-abelian gauge theories, if supplemented by assumptions
about the classification of the particles require universal "charges"
whose equality is imposed by the local gauge invariance of the theory.

In this talk we first collect the most important static properties of the lepton families (sec.2). In section 3 we examine the universality of coupling constants in abelian and non-abelian gauge theories. We discuss, in particular, the freedom in the choice of coupling constants that remains in the non-abelian case. Thus, testing universality means testing gauge theories and the classification of particles. In section 4 we review the present state of experimental tests of lepton unversality and point out what remains to be done. In the last section 5, finally, we comment briefly on reactions and decay processes that violate the conventional lepton number schemes.

2. The lepton generations and their main properties

The leptons occur in (at least) three generations f=e, μ, or τ,

$$
\begin{pmatrix} \nu_e \\ e^- \end{pmatrix} \qquad \begin{pmatrix} \nu_\mu \\ \mu^- \end{pmatrix} \qquad \begin{pmatrix} \nu_\tau^{(?)} \\ \tau^- \end{pmatrix} \tag{1}
$$

each of which seems to carry its own, additively conserved, lepton number L_f. The lepton number assignment could be as given in table 1. In any reaction involving leptons each of these lepton numbers seems to be conserved separately,

$$
\sum_i L_f(i) = \text{const.}, \quad f=e, \mu, \tau \tag{2}
$$

where i counts the leptons in the initial or the final state. The hint at these conserved lepton numbers comes from experimental evidence that ν_μ is not identical with ν_e, that the muon does not decay into an electron and a photon (the degree to which this is known is discussed below), and that ν_τ cannot be identical with either ν_μ, $\overline{\nu}_e$, or $\overline{\nu}_\tau$[+]. Thus,

$$
\nu_\mu \neq \nu_e; \quad \mu \nrightarrow e + \gamma; \quad \nu_\tau \neq \nu_\mu, \overline{\nu}_e, \overline{\nu}_\mu \tag{3}
$$

Another piece of evidence that is relevant for the question of lepton number assignments comes from a recent experiment at LAMPF[3]. This

---

[+] $\nu_\tau \equiv \nu_e$ cannot yet be excluded from experiment. That is the reason why we added a question mark in (1).

experiment tests for the decay process

$$\mu^+ \xrightarrow{?} e^+ + \bar{\nu}_e + \nu_\mu \qquad\qquad (A)$$

which is forbidden in the additive, sequential lepton number scheme. Thus, μ-decay should proceed in the following way,

$$\mu^+ \longrightarrow e^+ + \nu_e + \bar{\nu}_\mu \qquad\qquad (B)$$

and reaction (A) should not occur. The result of the LAMPF experiment for the decay rates is

$$\frac{R(A)}{(\mu \to all)} = -0.001 \pm 0.061$$

and confirms the conventional lepton number scheme.

In summary, up to date all experiments are compatible with this picture of sequential leptons, i.e. with the assumption that each lepton generation is characterized by its own, additively conserved lepton number.

The dynamical origin of these lepton numbers is not understood. Furthermore, it is not difficult to device gauge models in which these lepton numbers are not conserved individually (but their sum is) and yet which predict very small rates for processes of the kind indicated in eq. (3). Thus, it is very important to push the experimental upper bounds on these processes as low as possible and, at the same time, to reconsider lepton universality and to complete and to improve its experimental tests.

The fact that leptons come in three generations reveals a striking correspondence to the observation that the quarks also occur in three generations, viz.

$$\begin{pmatrix} u \\ d_C \end{pmatrix} \qquad \begin{pmatrix} c \\ s_C \end{pmatrix} \qquad \begin{pmatrix} t(?) \\ b \end{pmatrix} \qquad\qquad (4)$$

Here d_C and s_C denote the Cabibbo mixtures of down and strange quarks. (In fact, it seems that all three states, d, s, and b, are mixed).

This apparent quark-lepton symmetry and the fact that (at least) down and strange quarks appear as orthogonal mixtures in weak interactions has led to the speculation that the neutrino states that couple to weak interactions may be mixtures of ν_e and ν_μ (or of ν_e, ν_μ and ν_τ). If these states are not degenerate in mass this would provide a mechanism for violation of muonic (or tau-) lepton number, at a level below present experimental limits for processes of the type indicated in eq. (3). (See below, sec.5).

Table 2 summarizes our present knowledge of the <u>masses</u> of the leptons[4]. We note, in particular, the impressive accuracy to which the mass <u>ratio</u> m_μ/m_e is known. The situation regarding the neutrino masses is rather unsatisfactory: while we have a rather low upper limit on $m(\nu_e)$ from beta decay of the triton[5], the bounds on $m(\nu_\mu)$ and $m(\nu_\tau)$ are not very good. $m(\nu_\mu)$ is known to be smaller than about 650 keV/c^2 from the decay $K \rightarrow \pi\mu\nu_\mu$[6], or, less directly, to be smaller than about 500 keV/c^2 from a combination of the measurement of the muon momentum in pion decay at rest, $\pi \rightarrow \mu\nu_\mu$, and of the measured pion mass[7]. For ν_τ the present best upper limit comes from a measurement of the decay spectrum[8]. This measurement gives $m(\nu_\tau) < 250$ MeV/c^2. Beyond these laboratory experiments there are limits on the neutrino masses from astrophysical observation and cosmological models[9]. These limits depend, however, on additional information such as the lifetime of possible massive neutrino states. To quote just one example out of a somewhat complicated discussion of various possibilities: If the neutrinos are stable, and on the basis of the assumed theory of the universe, their masses should be either lower than about 45 eV/c^2 or higher than about 1 GeV/c^2 [+].

The issue of whether or not the neutrinos are massive is presently receiving great attention. Indeed, Reines et al. have reported positive evidence for neutrino oscillations[10], thus providing evidence for interfering neutrino states with a non zero mass difference. Furthermore a recent experiment has found evidence for a non-vanishing mass of the electron neutrino from triton beta decay[11]. If these results are

---

[+] The astrophysical limits, coming from rather indirect information, should probably not be taken too seriouly. On the contrary, any <u>direct</u> information on the neutrino masses is most welcome as an important input and boundary condition for cosmology.

confirmed and if it can be proved that ν_e is massive, it will be of even greater importance to improve on the mass limits for ν_μ and ν_τ or, more optimistically, to establish their masses. We note in passing that there are models in which the mass of neutrino f is proportional to the square of the mass of its charged partner,

$$m(\nu_f) = \text{const. } m_f^2$$

With a constant that may be only weakly dependent on (or even be com-pletely independent of) the lepton generation f[12]. With a ν_e mass of the order of some ten eV/c^2 one easily comes close to the present upper bounds on $m(\nu_\mu)$ and $m(\nu_\tau)$ ....

Regarding the interactions of leptons there is no indication for any <u>direct</u> coupling between lepton pairs of different generations. All interactions of leptons seem to be mediated by the photon γ and the weak vector bosons $W^\pm$, Z^0 etc.

More specifically,

(i) <u>Quantum Electrodynamics</u> of leptons is fully describable in terms of the universal interaction

$$\sum_f e_f \, j_{e.m.}^{(f)\alpha}(x) \, A_\alpha(x) \tag{6}$$

where $A_\alpha(x)$ is the quantized Maxwell vector potential, and $j_{e.m.}^{(f)\alpha}(x)$ is the electromagnetic current operator of lepton f. This current operator, of course, has the same form, as a functional of the field operators, for each lepton generation. The basic vertex representing the interaction (6) is drawn in fig. 1a.
The charge e_f is the same for all leptons,

$$e_f = e \quad \forall f$$

All quantitative differences in measurable quantities are solely due to the mass differences.

(ii) <u>Weak interactions via "charged currents"</u> between leptons (or bet-ween leptons and hadrons) is described by the exchange of positively and negatively charged W-bosons, as sketched in fig. 1b. These W-bosons $W_L^\pm$ couple to purely lefthanded currents of lepton fields. For small momentum transfer, i.e. $q^2 \ll m_W^2$, this gives rise to the well-known effective four-fermion interaction

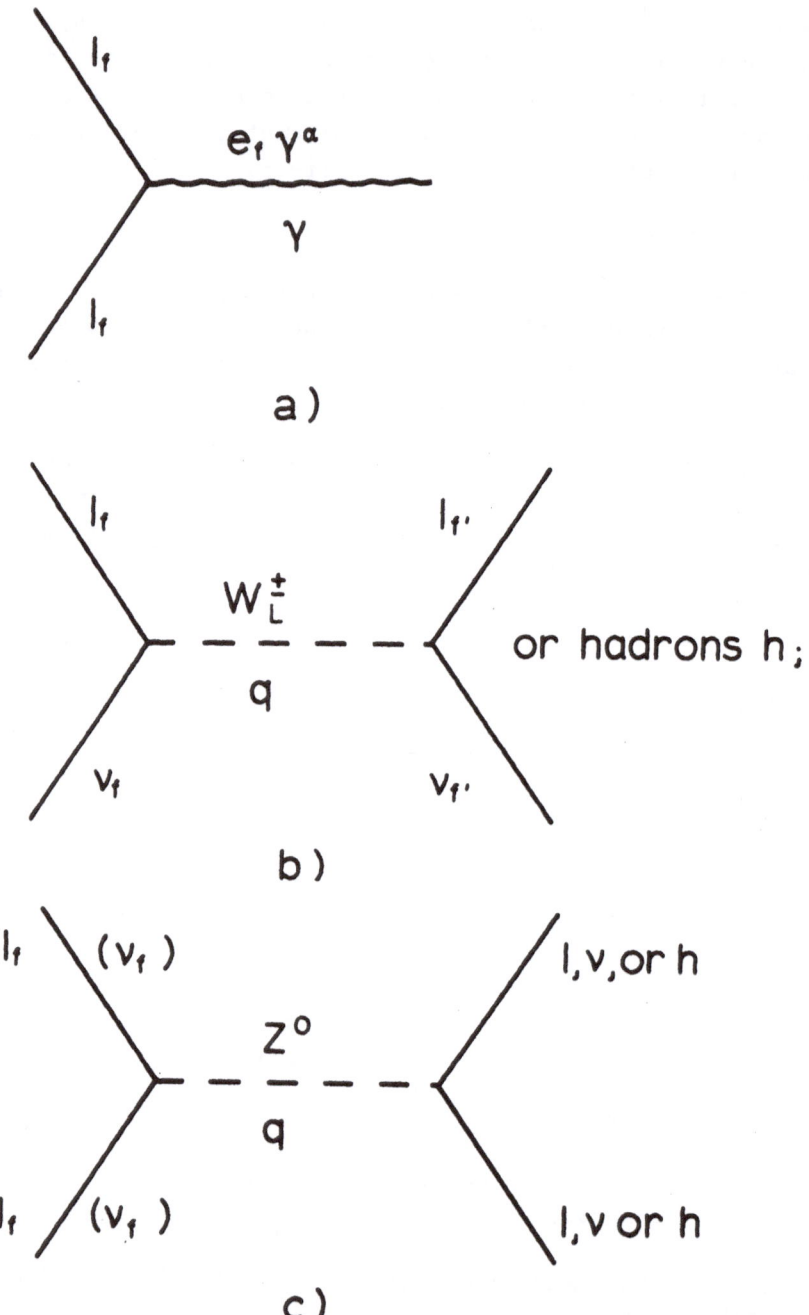

Fig. 1: These figures illustrate the electromagnetic, weak charged, and weak neutral interactions of leptons.

$$H^{eff}_{(\Delta Q=1)} \approx \frac{g^2}{8m_W^2} J_\alpha^+(x) J^\alpha(x) + h.c. \tag{7}$$

where

$$J_\alpha(x) = \sum_f \overline{\nu_f(x)} \gamma_\alpha (1-\gamma_5) l_f(x) + hadronic\ pieces \tag{8}$$

(iii) <u>Weak interactions via "neutral currents"</u> between leptons (and hadrons) is described by the exchange of at least one neutral vector boson Z^0. This is depicted in fig. 1c. This vector boson couples to the purely lefthanded current of neutrinos, but couples to a more complicated combination of vector and axial vector currents of their charged partners. Again, if $q^2 \ll m_{Z^0}^2$, an effective four-fermion interaction is obtained. For instance, in the unified theory of Weinberg and Salam we have

$$H^{eff}_{(\Delta Q=0)} \approx \frac{g^2}{16m_W^2} K_\alpha^+(x) K^\alpha(x) \tag{9}$$

where

$$K_\alpha(x) = \sum_f \overline{l_f(x)} \{(4\frac{m_Z^2-m_W^2}{m_Z^2} - 1)\gamma_\alpha + \gamma_\alpha\gamma_5\} l_f(x) +$$
$$+ \sum_f \overline{\nu_f(x)} \gamma_\alpha (1-\gamma_5) \nu_f(x) + hadronic\ pieces \tag{10}$$

In addition, the coupling constants e and g are universal and are related by

$$\frac{e^2}{g^2} = \frac{m_Z^2 - m_W^2}{m_Z^2} \equiv sin^2\theta_W \tag{11}$$

Here again, all quantitative differences in observables stem only from the difference in mass of the leptons involved. These lepton masses are neither predicted nor even related and must be put into the theory by hand.

We note, in passing, that at present all experiments on neutral currents, including those where also hadrons are involved, are well described by the Weinberg-Salam theory. The sine square of the Weinberg angle which is an open parameter, is found to be

$$sin^2\theta_W = 0.230 \pm 0.015 \tag{12}$$

This value leads to the prediction $m_W \simeq 78\ \text{GeV}/c^2$, $m_{Z^0} \simeq 89\ \text{GeV}/c^2$.

After this brief sketch of lepton properties and interactions we discuss the principle of universal coupling in somewhat more detail.

3. Universal coupling in the framework of gauge theories

Consider a theory of vector (gauge) fields $A_\alpha^{(n)}(x)$ and some fermion fields $\psi_f(x)$ defined by a Lagrange density invariant under a group G of local gauge transformations:

$$L = -\frac{1}{4}(F^{\mu\nu}, F_{\mu\nu}) + \sum_f (\overline{\psi_f(x)}\ (i\not{\partial} - m_f)\ \psi_f(x)) \tag{13}$$

Here $F^{\mu\nu}$ are the generalized field strength tensors pertaining to the gauge fields $A_\alpha^{(n)}(x)$. Their explicit form can be found in many articles on gauge theories[13] but is of no relevance here. $\not{\partial}$ is a shorthand notation for the group-covariant derivative,

$$\not{D}\ \psi_f(x) = (\not{\partial} - ig_f \sum_{n=1}^{N} U_f(T_n)\ A_\alpha^{(n)}(x)\ \gamma^\alpha)\ \psi_f(x) \tag{14}$$

T_n , for $n=1,2,\ldots,N$, are the generators of infinitesimal transformations of the gauge group G. $U_f(T_n)$ is the matrix representation of T_n in the space of the fields $\psi_f(x) = \{\psi_f^{(i)}(x);\ i=1,2,\ldots,M\}$ those fields spanning an M-dimensional (reducible or irreducible) representation of G.

The somewhat symbolical parentheses around the term $(F^{\mu\nu}, F_{\mu\nu})$, and, similarly, around the fermionic term in L , indicate that the fields should be coupled in a group invariant way.

For the sake of simplicity suppose that there are two different fermion generations $f=1,2$ in the theory. L shall be locally gauge invariant. The question that we ask is this: Can g_2 be different from g_1 , or, is the coupling of the fermion fields to the gauge fields universal or not?

Let us choose some local gauge transformation $\Lambda_1(x)$ on field $\psi_1(x)$, viz.

$$\psi_1'(x) = U_1\ (\Lambda_1(x))\ \psi_1(x)$$

The local gauge invariance of L , eq. (13), requires then that $\psi_2(x)$ be transformed too,

$$\psi_2'(x) = U_2 (\Lambda_2(x)) \psi_2(x)$$

and that the following two conditions on Λ_1 and Λ_2 be fulfilled:

$$g_1 (\partial_\mu U_2(\Lambda_2(x))) U_2^{-1}(\Lambda_2(x)) = g_2 (\partial_\mu U_2(\Lambda_1(x))) U_2^{-1}(\Lambda_1(x)) \quad (15)$$

$$U_2 (\Lambda_2^{-1}\Lambda_1) U_2(T_n) U_2^{-1}(\Lambda_2^{-1}\Lambda_1) = U_2(T_n) \quad (16)$$

It is not difficult to draw the relevant conclusions on g_1 and g_2 from these equations:

a)   If the group G is abelian, then

$$\Lambda_2^{-1}\Lambda_1 = \Lambda_1 - \Lambda_2 \qquad \text{and} \qquad U_f(T_n) = 1$$

Thus Λ_2 must be related to Λ_1 by

$$\Lambda_2(x) = \frac{g_2}{g_1}\Lambda_1(x)$$

Obviously, g_1 and g_2 can be chosen arbitrarily. Universality is not a consequence of local gauge invariance.

b)   If G is a non-abelian (simple or semi-simple) group then $\Lambda_2(x)$ must be the same as $\Lambda_1(x)$, and the coupling constants must be the same,

$$\Lambda_2(x) = \Lambda_1(x) ; \quad g_1 = g_2 \quad (18)$$

Thus, local gauge invariance enforces a universal coupling $g_f = g \Psi f$ for each irreducible component of the gauge group[+). However, this is not yet the whole story since g is not the actual "charge" through which the fermions couple to the physical vector fields. First of all, the __physical__ vector bosons eventually are linear combinations of the fields $A_\alpha^{(n)}(x)$. These combinations are determined by the specific mechanism of spontaneous symmetry breaking chosen (Higgs mechanism), i.e. by the Higgs sector to be added to L. Second, the actual coupling

---

[+)]   It is not essential that we considered fermion fields. The same reasoning applies to boson fields as well as to combinations of boson fields and fermion fields.

constant of any given fermion f to one of the physical vector bosons is given by g times some generalized magnetic quantum number (eigenvalue of some diagonal matrix $\sum_n a_n U_f(T_n)$).

Therefore, this charge depends on the representation spanned by the fields ψ_f. In summary, in the non-abelian case, there still is quite some freedom in the choice of coupling constants. In contrast to abelian gauge theory, where this freedom is continuous, in the non-abelian theory the charges can differ only by discrete steps. Universality, in the sense described in the last section, is only obtained if further assumptions about the classification of the fermion generations are made. In the Weinberg-Salam theory, in particular, the lepton generations are assumed to be classified in the same multiplets of $G \equiv SU(2) \times U(1)$. Each new generation is no more than a copy of the first.

So far, experimental tests of unified gauge theories of weak and electromagnetic interactions have been somewhat indirect. In particular, the data on weak interactions via neutral currents are in good agreement with the Weinberg-Salam theory but they do not exclude other unified gauge models. Direct tests such as the identification of the weak vector boson and - perhaps - the Higgs particles have to await the next generation of accelerators. In particular, the Higgs sector of the theory raises many questions to both theory and experiment that ought to be clarified.

In the meantime it seems to us of greatest importance to test lepton universality as precisely as possible. Testing universality means testing a fundamental building principle of unified gauge theories[†]. Furthermore, the nature of the lepton numbers needs to be clarified and experiments searching for violations of muon (or tau) lepton number must be pushed as far as possible.

[†] In fact, in the Higgs sector of gauge theories leptons of different mass have different couplings to the Higgs particles. These couplings are, however, well defined and are related among each other. In the Weinberg-Salam model the effects due to Higgs exchange, unfortunately, are very small.

4.   Experimental tests of lepton universality

There are many experiments on electromagnetic and weak interaction properties of leptons which test universality in a more or less direct way. We cannot give here a complete review of all such tests as this would go well beyond the scope of this talk. Instead we choose some characteristic examples that illustrate the state of the art and refer to the literature for a more complete survey.[1,2,14]

a)   Quantum Electrodynamics of electrons, muons and τ's provides many beautiful illustrations of the leptons' universal coupling to the quantized radiation field. In particular, muon-electron universality in QED results for the electromagnetic properties of free and bound muons are discussed in other talks at this symposion and we need not go into them. We choose only one example that seems to us particularly impressive: the g-factor anomaly of the muon. The anomaly which is defined as

$$a_\mu \equiv \tfrac{1}{2} (g_\mu - 2) \tag{19}$$

has been calculated from QED up to and including the order $O(\alpha^3)$. Some terms of order $O(\alpha^4)$ have been estimated and found to be small, of the order 4×10^{-9}. Fig. 2 shows the experimental value for a_μ, in comparison with its value $a_\mu(QED)$ computed on the basis of QED and μ-e universality. Also shown is a_μ as computed from QED and from higher order diagrams involving hadronic (and weak) virtual intermediate states such as the ones sketched in fig. 3. As can be seen the final theoretical value is in perfect agreement with the experimental value.

The hadronic contribution can be computed with good accuracy from an experimental input, namely the total cross section of e^+e^- into hadrons. The weak contribution is still small (about 2×10^{-9}). Thus, this comparison is a direct test of QED.

We find this result particularly beautiful and highly impressive for two reasons: (i) The prediction of universal QED of muons and electrons, using the physical masses of μ and e as input, is in perfect agreement with experiment. (ii) For this particular quantity one has tested QED up to its natural limit, i.e. to a precision where effects of the other interactions must be taken into account and, eventually, become overwhelming.

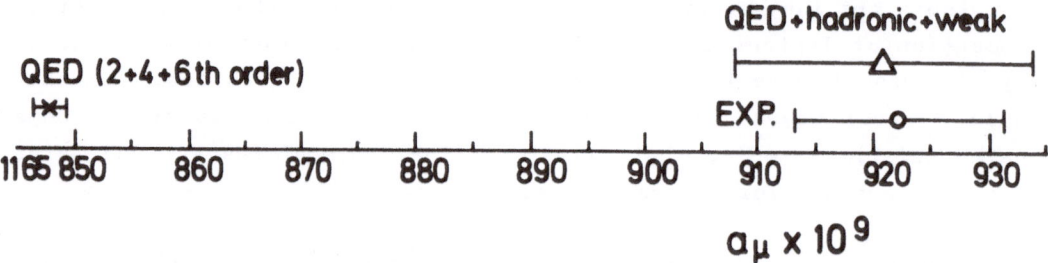

Fig. 2   Measured and predicted g-factor anomaly of the muon. The x and
         Δ are theoretical values, the open circle is the experimental
         result.

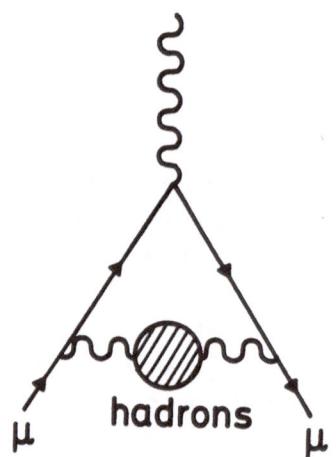

Fig. 3: Hadronic contributions to the g-factor anomaly of the muon via
        vacuum polarization through charged hadrons.

b)   Weak interactions via charged currents.

Here we quote three particularly illustrative examples: (i) If weak interactions of the "charged current" type are universal then the three processes

$$
\begin{aligned}
\pi^+ &\rightarrow e^+\nu_e \\
\pi^+ &\rightarrow \mu^+\nu_\mu \\
\tau^- &\rightarrow \pi^-\nu_\tau
\end{aligned}
\tag{20}
$$

are all proportional to the hadronic matrix element

$$
<o|a_\alpha^{(h)}(o)|\pi(q)> \quad = \quad \frac{i}{(2\pi)^{3/2}}\, f_\pi q_\alpha
\tag{21}
$$

(or its complex conjugate) with proportionality constants which contain G, the Fermi constant, and known functions of the lepton masses and the pion mass. In eq. (21) $a_\alpha^{(h)}$ denotes the weak hadronic axial vector current, q is the pion momentum $(q^2 = m_\pi^2)$, and f_π is the pion decay constant. Thus, in taking _ratios_ of the decay rates (20), the empirical constants f_π and G drop out and these ratios depend only on the lepton masses (up to radiative corrections).

In pion decay the experimental result is still the one published by Di Capua et. al.[15] sixteen years ago, (but corrected for the latest value of the pion lifetime as done in ref. 16),

$$
R(e/\mu) \equiv \frac{(\pi \rightarrow e\nu_e)}{(\pi \rightarrow \mu\nu_\mu)} = (1.274 \pm 0.024)\times 10^{-4}
\tag{22}
$$

Muon-electron universality, supplemented by radiative corrections, predicts

$$
R_{th}(e/\mu) = 1.233\times 10^{-4}
\tag{23}
$$

This is in fair agreement with the experimental result (22). Clearly, it is important to measure again $\pi \rightarrow e\nu_e$ decay and to improve on the result (22).

The prediction for the decay mode $\tau \rightarrow \pi\nu_\tau$ is also in fair agreement with experiment, as can be seen from table 3 below (third line), but the experimental uncertainty is still large.

(ii)   The purely leptonic decay processes

$$\mu^+ \rightarrow e^+\nu_e \overline{\nu_\mu} \tag{24a}$$

$$\tau^+ \rightarrow \mu^+\nu_\mu \overline{\nu_\tau} \tag{24b}$$

$$\tau^+ \rightarrow e^+\nu_e \overline{\nu_\tau} \tag{24c}$$

are completely predicted in the framework of the universal interaction (7). The decay rates, spectra, correlations and polarizations
are fully predictable in terms of G and of the lepton masses. For
example, the spectrum parameter ρ (Michel parameter) in μ-decay (24a)
is found to be

$$\rho(\mu) = 0.752 \pm 0.003 \tag{25}$$

whilst in the τ-decay modes (24b) and (24c) it is found to be

$$\rho(\tau) = 0.72 \pm 0.15 \tag{26}$$

The values (25) and (26) agree with each other, within the error bars,
and compare well with the theortical "V-A" value $\rho=3/4$.

(iii)   The τ lepton was discovered at a time where charged weak interaction was already well understood. Thus, assuming lepton universality,
all decay modes of the τ could be computed in advance[17]. All that remained to be done after its discovery was to insert its mass into the
calculated rates. Table 3 summarizes the present state of knowledge on
the τ-decays[2]. Even though the error bars are still large, the overall
agreement of the data with the predictions on the basis of lepton universality is impressive.

c)   Weak interactions via neutral currents

Here the following processes have been measured, testing lepton
universality in the neutral weak interactions (as predicted by eqs. (9)
and (10):

(i)   elastic scattering of neutrinos on electrons,

$$\nu_\mu + e \rightarrow \nu_\mu + e \tag{27a}$$

$$\overline{\nu_\mu} + e \rightarrow \overline{\nu_\mu} + e \tag{27b}$$

$$\overline{\nu_e} + e \rightarrow \overline{\nu_e} + e \tag{27c}$$

(In the third process (27c) there is also a contribution from charged currents. This contribution is subtracted out in the comparison shown below, fig. 4.)

(ii) Inclusive scattering of left-handed and right-handed electrons on deuterium, and parity violating effects in electronic atoms.

We cannot enter into this second rather vast topic, for lack of time. However, as the basic arguments are simple here, we may sketch briefly the results for the reactions (27). We assume that neutrinos are fully left-handed (and, therefore, antineutrinos are fully right-handed), (assumption I). The neutral currents of electrons and neutrinos shall be, respectively,

$$\overline{e(x)} \{C_V \gamma_\alpha + C_A \gamma_\alpha \gamma_5\} e(x) \tag{28a}$$

$$\overline{\nu(x)} \gamma_\alpha (1+\lambda\gamma_5) \nu(x) \tag{28b}$$

It is not difficult to compute the total cross sections for ν and $\overline{\nu}$ scattering on the electron. For $E_\nu^{lab} \gg m_e$ (E_ν^{lab} is the energy of the neutrino in the laboratory system) we find

$$\sigma_{tot}\left(\left[\begin{matrix}\nu\\\overline{\nu}\end{matrix}\right]e\right) \simeq \frac{2m_e G^2}{\pi} E_\nu^{lab} \frac{|1-\lambda|^2}{4} \cdot \frac{1}{3}\{|C_V|^2+|C_A|^2\left[\begin{matrix}-\\+\end{matrix}\right]Re(C_V^* C_A)\} \tag{29}$$

where

$$\frac{2m_e G}{\pi} = 1.6 \times 10^{-41} \text{ cm}^2/\text{Gev} \tag{30}$$

We believe that λ is equal to -1, see eq. (10), but this is of no importance (as long as $\lambda \neq +1$). Let us further assume that C_V and C_A are relatively real (assumption II). Then the expression in curly brackets becomes

$$\{C_V^2 + C_A^2 \left[\begin{matrix}-\\+\end{matrix}\right] C_V C_A\} \tag{31}$$

This is an ellipse in the plane (C_V, C_A) whose symmetry axes are rotated by $\pm\frac{\pi}{4}$ with respect to the coordinate axes. Thus, if one plots the total cross sections divided by the neutrino energy, the three ellipses corresponding to the processes (27) should intersect in the same two points. These intersection points determine C_V and C_A, up to a twofold ambiguity.

Fig. 4 shows this comparison with the measured cross sections for processes (27)[+].

Needless to add that one of the intersection points agrees with the Weinberg-Salam theory. It must be stressed, however, that this comparison can only be made if assumptions (I) and (II) are made. In particular, assumption (I) is essential: In principle, total cross sections (very much like decay rates) are not correlation observables odd under parity and, therefore, cannot give information on relative sign and magnitude of axial vector and vector interactions. Only spin-momentum correlations can do that. In this specific case the spin-momentum correlation is introduced through the assumption of left-handed neutrinos, whilst phase ambiguities are eliminated through the assumption that C_A and C_V are relatively real.

Our knowledge of neutral weak interaction, and, therefore, of lepton universality in this sector, is incomplete. Experiments that are being prepared or that seem feasible in an immediate future comprise
(i) Scattering of left-handed and right-handed electrons on nuclei, both inclusive and elastic[19]. (Measurement of parity violating asymmetry).
(ii) The analogue experiments with high-energy polarized muons, measuring a parity violating asymmetry in the scattering on nucleons and nuclei.
(iii) Measurement of parity violating effects in muonic atoms[20].

The last two groups of experiments are of special importance since nothing is known, as yet, about μ-e or τ-μ-e universality in neutral weak interactions. Obviously this is a crucial test of lepton universality in general, and of the predictions of unified gauge theories in particular.

5.   Processes violating additive lepton number conservation

We close this talk with a few comments on decay processes and reactions which test the additive conservation of muon (and tau) lepton numbers. Obviously, the nature of lepton numbers and their conservation laws is an essential problem both in connection with lepton

---

[+] The data on the processes (27) and a complete set of refences can be found in ref. 18. Fig. 4 was kindly provided to me by Helmut Faissner.

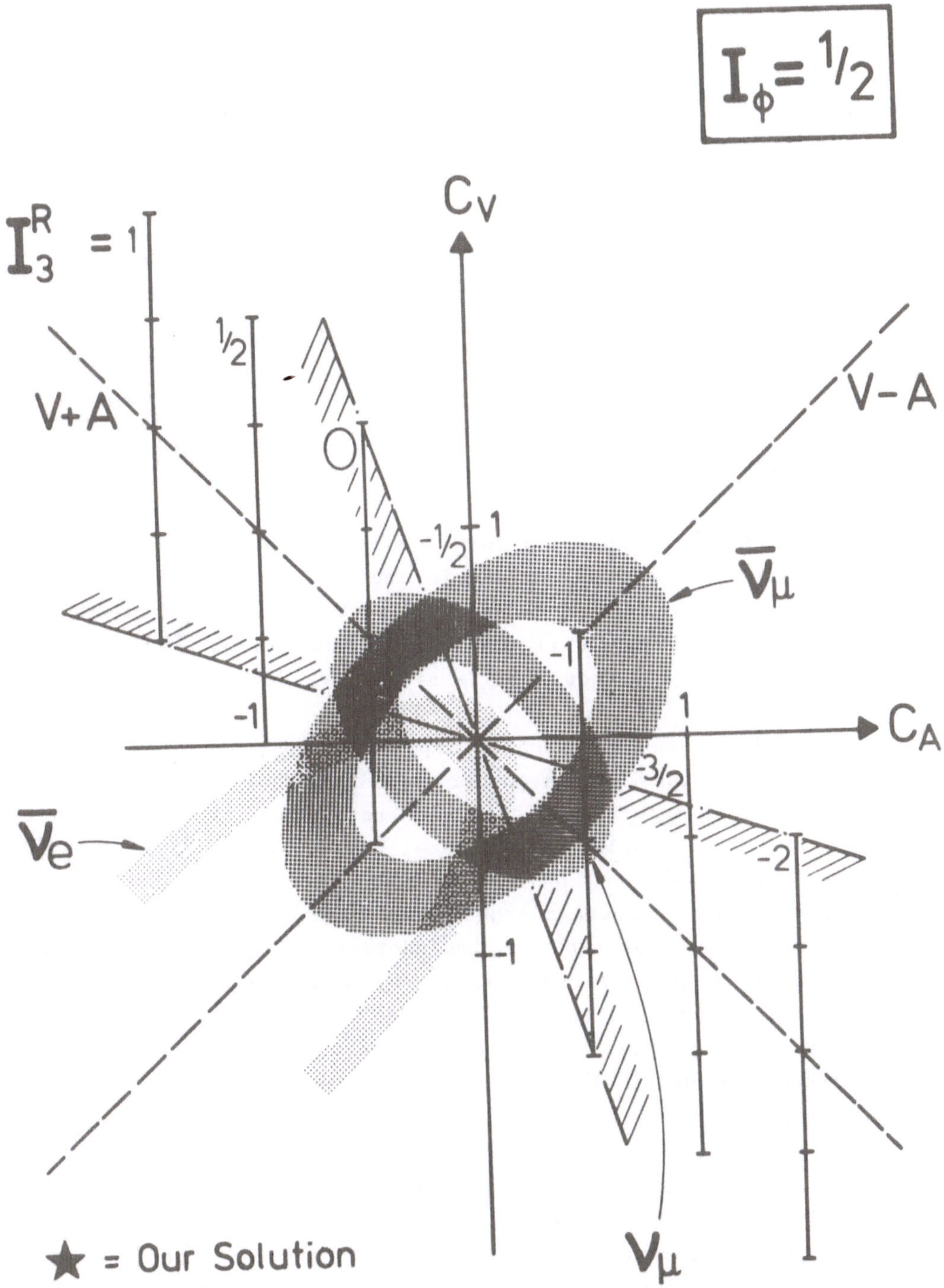

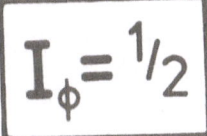

Fig. 4: Bounds on V and A coupling constants provided by $\nu_\mu e$, $\bar{\nu}_\mu e$, and $\bar{\nu}_e e$ total cross sections. Vertical bars indicate allowed range of x_W' for different I_3-assignments to right-handed electron, and Higgs isospin $I_\phi = 1/2$. Ratio of electron energies admits sectors around V + A axis.
(All errors ± 1 s.d.)

universality as postulated by unified gauge theories, as well as in
connection with the general observation that the leptons come in
generations.

Processes of interest, in this respect, are

$$\mu^+ \;\rightarrow\; e^+\gamma \tag{32}$$

$$\mu^- \;\rightarrow\; e^- \quad\quad ; \text{ neutrinoless conversion on a nucleus} \tag{33}$$

$$\mu^- \;\rightarrow\; e^+ \quad\quad ; \text{ neutrinoless conversion on a nucleus} \tag{34}$$

$$\nu_\mu \;\rightarrow\; \nu_e\gamma \tag{35}$$

$$\tau^+ \;\rightarrow\; \mu^+\gamma \tag{36}$$

$$\tau^+ \;\rightarrow\; e^+\gamma \tag{37}$$

etc. All of these processes are forbidden if the additive lepton number
scheme applies, as we described it in sec.2. However, if only the sum
of L_e , L_μ , L_τ is conserved but the individual lepton numbers are
not, then reactions (32), (33), (35), (36), and (37) are allowed. μ^- to
e^+ conversion, eq. (34), however, is still forbidden.

On the other hand, if some other, more complicated scheme applies
which assigns the same lepton number to the particle of one generation
and to the antiparticle of another, then reaction (34) could be allowed
while all the others would be forbidden. Even if this happened we would
expect $(\mu^- e^+)$-conversion on a nucleus to be very improbable because
this process needs a double charge exchange in a weak reaction, as two
protons have to be converted into two neutrons. This can only happen by
means of two weak boson exchanges, or, by means of one weak boson and
one pion exchange on a virtual π^+ , or Δ^+ , or Δ^{++} . In any case,
the capture rate is likely to come out very small. So any such more
exotic scheme is difficult to test with muons, taus, electrons, and
nucleon targets.

The processes (32) to (34), (as well as $\mu\rightarrow e\bar{e}e$ which we do not
discuss here), have been studied extensively, both experimentally and
theoretically. None of them has been seen as yet; the recent generation
of experiments have lowered the upper limits for the branching ratios
to an impressive level of sophistication. The latest results are these:

$$R_{\mu e\gamma} \; := \; \frac{\Gamma(\mu \rightarrow e\gamma)}{\Gamma(\mu \rightarrow \text{all})} \; < \; 1.9 \times 10^{-10} \qquad (\text{ref. 21})$$

$$R_{\mu e} \; := \; \frac{\Gamma(\mu^- + (A,Z) \rightarrow (A,Z)^* + e^-)}{\Gamma(\mu^- \rightarrow \nu_\mu;\ \text{capture})} \; < \; 7 \times 10^{-11}$$

(This result was obtained on Sulfur, see ref. 22)

$$R_{\mu\bar{e}} \; := \; \frac{\Gamma(\mu^- + (A,Z) \rightarrow (A,Z-2)^* + e^+)}{\Gamma(\mu\ \text{capture})}$$

$$R_{\mu\bar{e}} \; < \; 9 \times 10^{-10} \qquad (\text{on S, ref. 22});$$

$$R_{\mu\bar{e}} \; < \; 3 \times 10^{-10} \qquad (\text{on } {}^{127}\text{I, ref. 23})$$

Various groups at LAMPF, SIN and TRIUMF are planning further experiments on these processes and on the related process $\mu^+ \rightarrow e^+ e^- e^+$, and expect to reach the level of 10^{-12} in the next step.

What are the theoretical possibilities of introducing breakdown of lepton number conservation, in the framework of gauge theories, and what are typical predictions of such extended models ?

It is not difficult to break muon (and/or tau) lepton number in unified gauge theories[24,25]. For example, assume that the mass eigenstates "ν_e" and "ν_μ" have a nonvanishing difference in mass. Assume further that the neutral partners of electron and muon in left-handed weak interactions are orthogonal mixtures of these states, viz.

$$\begin{pmatrix} \nu'_e \\ e \end{pmatrix}_L \; ; \qquad \begin{pmatrix} \nu'_\mu \\ \mu \end{pmatrix}_L \qquad \text{with} \qquad \begin{aligned} \nu'_e &= \nu_e \cos\delta + \nu_\mu \sin\delta \\ \nu'_\mu &= -\nu_e \sin\delta + \nu_\mu \cos\delta \end{aligned} \qquad (38)$$

In this minimal model the rate for $\mu \rightarrow e\gamma$ is finite and non-zero. One finds

$$R_{\mu e\gamma} \; = \; \frac{75\alpha}{128\pi} \sin^2(2\delta) \left(\frac{m^2(\nu_\mu) - m^2(\nu_e)}{m_W^2} \right)^2 \qquad (39)$$

Unfortunately this rate is very small even if the mixing is optimal. Take for instance the following set of parameters:

$$m(\nu_\mu) = m_e \; ; \; m(\nu_e) = 0 \; ; \; m_W = 78\,\text{GeV}/c^2 \; , \; \text{and} \; \delta = \frac{\pi}{4}$$

This gives $\qquad R_{\mu e\gamma} \; \approx \; 2.5 \times 10^{-24}$

Thus, in this minimal extension of the Weinberg-Salam or any other unified gauge theory, there could well be strong muon number violation without this being measurable in the processes (32) and (33). On the other hand, one can make the theoretical prediction much larger if new heavy neutral (and, with some care, also charged) leptons are introduced in the place of ν_μ and ν_e which are too light for that purpose. Of course, these particles must be endowed with reasonably strong couplings to e , μ , $W^\pm$, and Z^0. The calculated rate then depends on these unknown coupling constants and on unknown mass differences, and, therefore, is not a genuine prediction any more. The only result of these considerations which seems to be somewhat less model dependent, is the statement that the branching ratio for (μe) conversion is generally larger, by up to two orders of magnitude, than $R_{\mu e \gamma}$ [25].

In summary, we may say this: Unified gauge theories, although they may break muon and/or tau lepton number, have very little predictive power as to the rates for all $\mu \rightarrow e$ processes. The question of whether or not $L_\mu (L_\tau)$ is an exactly conserved quantity is completely open: it could be that L_μ is broken but that $\mu \rightarrow e \gamma$ and (μe)-conversion are not seen, at the level investigated so far, because they are dynamically suppressed. With this unsatisfactory state of theory any experimental progress on these ultrarare processes is of utmost importance and will be most welcome.

References:

1. F. Scheck; "Muon Physics", Phys. Rep. 44 (1978) 187

2. G. Flügge, "The New Heavy Lepton τ", Z. Physik C1 (1979) 121

3. S.E. Willis et al.; Phys. Rev. Letters 44 (1980) 522, Errata: ibid. 45 (1980) 1370

4. Review of Particle Properties, Rev. of Mod. Phys. 52, No.2 (1980)

5. List of references can be found in 4.

6. A.R. Clark et al.; Phys. Rev. D9 (1974) 533

7. M. Daum et al.; Phys. Lett. 60B (1976) 380; Phys. Rev. D20 (1979) 2692

8. W. Bacino et al.; Phys. Rev. Letters 42 (1979) 749

9. R. Cowsik and J. McClelland; Phys. Rev. Lett. 29 (1972) 669
    G. Steigmann et al.; Phys. Letters 66B (1977) 165
    B.W. Lee and S. Weinberg; Phys. Rev. Letters 39 (1977) 165
    D.A. Dicus et al.; Phys. Rev. Letters 39 (1977) 168
    K. Sato et al.; Progr. Theor. Phys. 58 (1977) 1775
    See also N. Straumann, "Neutrinos and Cosmology"; Proceedings of SIN Spring School, Zuoz 1978

10. F. Reines et al.; Phys. Rev. Letters 45 (1980) 1307

11. V.A. Lubimov et al.; Phys. Letters 94B (1980) 266

12. See e.g. R.N. Mohapatra and G. Senjanović; Phys. Rev. Letters 44 (1980) 912

13. See e.g. the excellent review by L. O'Raifeartaigh, Rep. Prog. Phys. 42 (1979) 159

14. M.M. Nagels et al.; Nucl. Phys. B109 (1976) 1

15. E. Di Capua et al.; Phys. Rev. 133 (1964) B 1333

16. D. Bryman and C. Picciotto; Phys. Rev. D11 (1975) 1337

17. Y.S. Tsai; Phys. Rev. D4 (1971) 2821
    H.B. Thacker and J.J. Sakurai; Phys. Letters 36B (1971) 103
    J.D. Bjorken and C.H. Llewellyn-Smith; Phys. Rev. D7 (1973) 887

18. B.C. Barish; Phys. Rep. 39C (1978) 279

19. M Fischer-Waetzmann and F. Scheck; Phys. Rev. D21 (1980) 2510 and references quoted therein.

20. J.H. Missimer and L. Simons; Nucl. Phys. A316 (1979) 413 and further references therein.

21. J.D. Bowman et al.; Phys. Rev. Letters 42 (1979) 556

22. A. Badertscher et al.; Phys. Letters 79B (1978) 371

23. R. Abela et al.; Preprint Basel University (1980)

24. T.P. Cheng and L.F. Li; Phys. Rev. D16 (1977) 1425

25. W.J. Marciano and A.I. Sanda; Phys. Rev. Letters 38 (1977) 1512 and Phys. Letters 67B (1977) 303

	L_e	L_μ	L_τ
$e^-\ \nu_e$	1	0	0
$e^+\ \bar\nu_e$	-1	0	0
$\mu^-\ \nu_\mu$	0	1	0
$\mu^+\ \bar\nu_\mu$	0	-1	0
$\tau^-\ \nu_\tau$	0	0	1
$\tau^+\ \bar\nu_\tau$	0	0	-1

Table 1: The three lepton generations and their lepton number assignment in the scheme of "sequential leptons".

Lepton	Mass	Remarks
e	$0.5110034(14)\,\mathrm{MeV/c^2}$	
ν_e	$< 60\ \mathrm{eV/c^2}$	from $^3H \rightarrow {}^3He + e^- + \nu_e$
	$14\mathrm{eV} \leq m(\bar\nu_e) \leq 46\mathrm{eV}$	See ref. 11
μ	$105.65946(24)\ \mathrm{MeV/c^2}$	$m_\mu/m_e = 206.76859(29)$ (1.4 ppm)
ν_μ	$< 650\ \mathrm{keV/c^2}$	from $K^+ \rightarrow \pi^0\mu^+\nu_\mu$
	$< 500\ \mathrm{keV/c^2}$	from $|\vec{p}_\mu|$ in pion decay at rest
τ	$1782^{+3}_{-4}\ \mathrm{MeV/c^2}$	
ν_τ	$< 250\ \mathrm{MeV/c^2}$	from decay spectrum

Table 2. Lepton masses and bounds on neutrino masses

Decay mode	prediction	experiment
$\tau^- \rightarrow e^- \bar{\nu}_e \nu_\tau$	16.8 %	17.1 ± 1.0 %
$\tau^- \rightarrow \mu^- \bar{\nu}_\mu \nu_\tau$	16.4 %	17.5 ± 1.2 %
$\tau^- \rightarrow \pi^- \nu_\tau$	9.5 %	9.8 ± 1.4 %
$\tau^- \rightarrow \rho^- \nu_\tau$	25.3 %	20.5 ± 4.1 %
$\tau^- \rightarrow A_1^- \nu_\tau$	8.1 %	10.4 ± 2.4 %
$\tau^- \rightarrow \nu_\tau \rightarrow$ 3 charged particles	$\sim$ 26 %	30.6 ± 3.0 %

Table 3. Predicted and measured decay modes of the τ lepton.

TEST OF QUANTUM ELECTRODYNAMICS AT HIGH MOMENTUM TRANSFERS

V. Hepp[*]

Universität Hamburg

1. Introduction

With the advent of the e^+e^- colliding beam facility PETRA at DESY (Hamburg)
tests of the validity of Quantum Electrodynamics (QED) at very large momentum
transfers $|q^2|$ have become possible. As is well known, no deviations from QED
have been seen at lower energies[1].

Precise tests of QED at high $|q^2|$ are of great importance both for atomic and
for high energy physics. A determination of a cut-off mass Λ of ~ 100 GeV may
be translated into a distance $\lambda = \hbar c/\Lambda$ ~ $2 \cdot 10^{-16}$ cm which is a measure of the
point-like structure of the electron. High precision atomic physics experiments
(e.g. measurements of the anomalous magnetic moment of the leptons) depend
strongly on the lower bound of Λ when compared to theory. High energy experiments with colliding beams rely heavily on the validity of QED, since all cross
sections for new phenomena are normalized to Bhabha scattering. Above all,
tests of QED are of fundamental importance, because QED is the first successful
gauge theory.

In table 1 a summary of existing and planned e^+e^- storage rings is given. At
the moment the PETRA machine has attained the highest CMS energy ($\sqrt{s} = 2\ E_{beam}$
~ 35 GeV, corresponding to $|q^2|_{max}$ ~ 1225 GeV²). The PEP ring at Stanford is
just starting to work at similar energies. The planned European LEP project is
designed to achieve beam energies of up to 130 GeV, corresponding to $|q^2|_{max}$ ~
67600 GeV². For comparison, previous QED tests at SPEAR[2] covered momentum
transfers up to ~ 50 GeV².

In this talk QED tests at PETRA are presented. The following reactions have
been studied:

----

[*] On leave of absence from University of Heidelberg

- Bhabha scattering: $e^+e^- \rightarrow e^+e^-$ (1)
- Lepton-pair production: $e^+e^- \rightarrow \mu^+\mu^-, \tau^+\tau^-$ (2)
- Two-photon annihilation: $e^+e^- \rightarrow \gamma\gamma$ (3)

Data were taken with the detectors JADE[3], MARK J[4], PLUTO[5] and TASSO[6] at CMS energies up to ~ 35 GeV. For details of the apparatus and of the experimental procedure we refer to the references. A discussion of the two-photon exchange process $e^+e^- \rightarrow e^+e^-$ + lepton pair which occurs at predominantly low q^2[7] will be omitted in this presentation.

The talk is structured as follows: in section 2 possible modifications of QED are outlined and their effects on the cross sections for reactions (1) to (3) are discussed. Experimental results are given in section 3 and the already sizeable effect of electro-weak interference is discussed in section 4. The last section 5 gives a summary.

2. Modifications of QED

The differential cross sections for reactions (1) to (3) may be written in the form:

$$\frac{d\sigma}{d\Omega} = \frac{d\sigma_0}{d\Omega}(1 + \delta_{rad} + \delta_{had} + \delta_w + \delta_\Lambda) = \frac{d\sigma_{QED}}{d\Omega}(1 + \delta_w + \delta_\Lambda) \qquad (4)$$

where $\frac{d\sigma_0}{d\Omega}$ denotes the lowest order QED cross section; $\frac{d\sigma_{QED}}{d\Omega}$ incorporates radiative corrections (δ_{rad}) and hadronic vacuum polarisation (δ_{had}).

The expected correction δ_w due to electro-weak interference is small at present (see section 4). Any further deviation from the known theory may be incorporated into a correction δ_Λ which is in general a function of scattering angle θ and CMS energy $\sqrt{s}$. The actual parameterisation of δ_Λ will be discussed in the following.

QED modifications are expected if heavy photon-like objects (γ^*, neutral Higgs bosons etc.) or massive leptons (E^*) exist with finite coupling strengths.

Their presence will modify both the lepton-photon vertices, e.g.

and the photon and lepton propagators, e.g.

or

The differential cross sections will be affected in a different way for diagrams involving inner photon lines (Bhabha scattering and lepton pair production) than for charged lepton exchange (two-photon annihilation).

2.1. Real leptons, virtual photons: $e^+e^- \to \ell^+\ell^-$

If we take Bhabha scattering as an example, two types of diagrams contribute to the lowest order cross section:

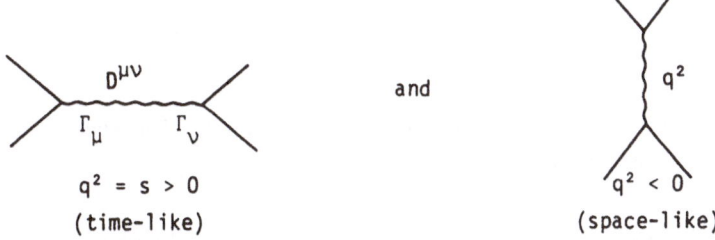

and

$D^{\mu\nu}$

Γ_μ Γ_ν

$q^2 = s > 0$
(time-like)

q^2

$q^2 < 0$
(space-like)

The matrix element M_{if} is built up from the vertex functions $\Gamma_\mu = \gamma_\mu F(q^2) +$ {small $\sigma^{\mu\nu} q^\nu$ term*} and the photon propagator $D^{\mu\nu} = -D(q^2) g^{\mu\nu}/q^2$. The form factor $F(q^2)$ and $D(q^2)$ are of course unity in standard QED. The differential cross section measures the product $|F^2 D|^2$.

Deviations from QED can arise from a modification of the photon propagator $D(q^2)$ for $q^2 \neq 0$, e.g. by a heavy photon, which may lead to the replacement

$$\frac{1}{q^2} \to \frac{1}{q^2} + \frac{1}{q^2 - \Lambda_p^2} \approx \frac{1}{q^2} (1 + \frac{q^2}{\Lambda_p^2}) \qquad \text{for } q^2/\Lambda_p^2 \ll 1 \qquad (5)$$

---

* This term is of the order of ~ 0.2 % in standard QED.

In the static limit this corresponds to a change of the r dependence of the Coulomb potential:

$$\frac{1}{r} \rightarrow \frac{1}{r}(1 - e^{-\Lambda_p r}) \ . \tag{6}$$

Similarly such a heavy photon will modify the vertex $F(q^2)$ for $q^2 \neq 0$ for which we again keep only the first term in the expansion with respect to q^2:

$$\Gamma_\mu = \gamma_\mu F(q^2) \approx \gamma_\mu(1 + q^2/\Lambda_V^2) \tag{7}$$

Both modifications are usually parameterized with common form factors F_s, F_t, allowing for a difference in the space-like (subscript s) or time-like (subscript t) region of four momentum transfer either by pole terms motivated by eq. (5) or just by the linear expansion:

$$\text{time-like:} \quad F_t = F^2 D = 1 \mp \frac{s}{s - \Lambda_\pm^2} \approx 1 \pm \frac{s}{\Lambda_\pm^2}$$

$$\text{space-like:} \quad F_s = F^2 D = 1 \mp \frac{q^2}{q^2 - \Lambda_\pm^2} \approx 1 \pm \frac{q^2}{\Lambda_\pm^2} \tag{8}$$

Note that $\Lambda_\pm$ describes globally the effect due to Λ_V or Λ_p. The subscript $\pm$ refers to the sign of the correction.

Hence we obtain for the modified Bhabha cross section ($q^2 = -s \sin^2\frac{\theta}{2}$; $q'^2 = -s \cos^2\frac{\theta}{2}$):

$$\frac{d\sigma}{d\Omega} = \frac{\alpha^2}{2 s} \left\{ \frac{q'^4 + s^2}{q^4} |F_s|^2 + \frac{2 q'^4}{q^2 s} \text{Re}(F_s F_t^*) + \frac{q'^4 + q^4}{s^2} |F_t|^2 \right\} \ .$$

$$\cdot (1 + \delta_{rad} + \delta_{had}) \tag{9}$$

or

$$\frac{d\sigma}{d\Omega} = \frac{d\sigma_{QED}}{d\Omega} (1 + \delta_\Lambda(s,\theta)) \quad \text{with } \delta_\Lambda(s,\theta) \approx \pm \frac{3 s}{\Lambda_\pm^2} \frac{1 - \cos^2\theta}{3 + \cos^2\theta}$$

For $\Lambda = 100$ GeV, $\sqrt{s} = 31$ GeV: $\delta_\Lambda \sim 10 \%$ at 90° and zero at 0°.

The $\mu^+\mu^-$ and $\tau^+\tau^-$ cross section which has only contributions from time-like photons is modified as ($m_\tau \ll \sqrt{s}$):

$$\frac{d\sigma}{d\Omega} = \frac{\alpha^2}{4s} (1 + \cos^2\theta) \, |F_t|^2 \, (1 + \delta_{rad} + \delta_{had}) \tag{10}$$

$$= \frac{d\sigma_{QED}}{d\Omega} (1 + \delta_\Lambda) \quad \text{with} \quad \delta_\Lambda \sim - \frac{2s}{\Lambda_\pm^2}$$

In this case δ_Λ is independent of θ; for $\Lambda = 100$ GeV, $\sqrt{s} = 31$ GeV: $\delta_\Lambda \sim 20\ \%$.

2.2. Virtual leptons, real photons: $e^+e^- \rightarrow \gamma\gamma$

Here the QED Feynman diagrams are:

where p, q, k are the four momenta of real and virtual lepton and real photon, respectively.

For the two-photon annihilation QED modifications will show up only in $O\ (q^4/\Lambda^4)$. This is true both for the "sea-gull" and the heavy electron (E^*) graph and needs some explanation.

(a) "Sea-gull" graph

We shall illustrate the cancellation of the $O\ (q^2/\Lambda^2)$ terms by looking at a simple set of diagrams:

The dashed lines symbolize the coupling of a new neutral object to the fermions. In this case gauge invariance gives an important constraint to the q^2 dependence of the modification. The Ward-Takahashi Identity[8] relates the divergence of e.g. the upper vertex function to the fermion propagators S_F:

$$k^\mu \; \Gamma_\mu(q) \;\; = \;\; S_F^{-1}(p) \; - \; S_F^{-1}(q) \tag{11}$$

For one momentum on-shell ($\not{p} = m;\; S_F^{-1}(p) = 0$) it may be written as:

$$k^\mu \; \Gamma_\mu(q) \; S_F(q) \;\; = \;\; -1$$

If we approximate $\quad \Gamma_\mu(q) \;\; = \;\; \gamma_\mu \; F(q^2) \quad$ we obtain:

$$k^\mu \; \gamma_\mu \; F(q^2) \; S_F(q) \;\; = \;\; -1 \quad \text{ or } \quad F(q^2) \; S_F(q) \;\; = \;\; \frac{1}{\not{q} - m} \quad ,$$

$$\qquad\qquad\qquad\qquad\qquad\qquad\qquad\qquad \text{"dressed"} \qquad \text{"naked"}$$

i. e. the modification cancels completely for one vertex and the adjacent propagator.

For the second vertex there are still cancellations against the "two-photon vertex" (see above figure). It can be shown[9,10] that all corrections $0 \; (q^2/\Lambda^2)$ cancel and that the modification can be parameterized by form factors:

$$F(q^2) \;\; \sim \;\; 1 \pm \frac{q^4}{\Lambda_\pm^4}$$

The Ward-Takahashi-identity gives no restriction on the $\sigma_{\mu\nu} \; k^\nu$ term which effectively contributes also only in $0 \; (q^4/\Lambda^4)$ (see (b)).

Hence:

$$\frac{d\sigma}{d\Omega} \;\; = \;\; \frac{\alpha^2}{2s} \; \{ \; \frac{q'^2}{q^2} \; |F(q^2)|^2 \; + \; \frac{q^2}{q'^2} \; |F(q'^2)|^2 \} \; (1 + \delta_{rad})$$

$$\;\; = \;\; \frac{d\sigma_{QED}}{d\Omega} \; (1 + \delta_\Lambda) \quad \text{ with } \quad \delta_{\Lambda_\pm} \; \sim \; \pm \; \frac{s^2}{2 \; \Lambda_\pm^4} \; \frac{\sin^4\theta}{1 + \cos^2\theta} \tag{12}$$

(b) Heavy electron (E^*) exchange

Consider the interference of:

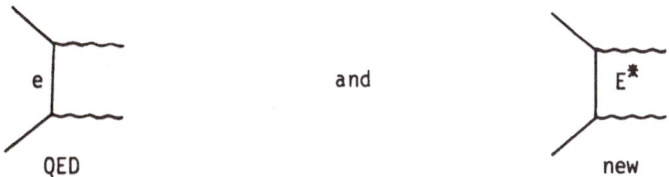

QED   and   new

Current conservation excludes a γ_μ coupling between e, E^* and γ. The allowed magnetic moment coupling leads for dimensional reasons to a matrix element $M_{if}(E^*)$ of the rough form

$$M_{if}(E^*) \sim \frac{e^{*2}}{\Lambda^2} \cdot \frac{s}{q^2 - \Lambda^2}$$

where we have written the magnetic transition moment as e^*/Λ.
The QED-matrix element $M_{if}(e)$ is proportional to $e^2/(q^2 - m_e^2)$.
Hence we expect qualitatively for $m_e^2 \ll q^2 \ll \Lambda^2$:

$$\frac{M_{if}(E^*)}{M_{if}(e)} \sim \frac{e^{*2}}{e^2} \cdot \frac{s\,q^2}{\Lambda^4} \sim \frac{\delta_\Lambda}{2} \sim s^2 \cdot \sin^2\theta$$

Here Λ signifies the mass of the heavy electron E^* if the coupling e^* is equal to e. A quantitative calculation of the cross section[11] yields:

$$\frac{d\sigma}{d\Omega} = \frac{\alpha^2}{s} \cdot \frac{1 + \cos^2\theta}{1 - \cos^2\theta} \cdot (1 + \frac{s^2}{2\,\Lambda^4} \sin^2\theta) \cdot (1 + \delta_{rad}) \tag{13}$$

$$= \frac{d\sigma_{QED}}{d\Omega}(1 + \delta_\Lambda) \quad \text{with} \quad \delta_\Lambda = \frac{s^2}{2\,\Lambda^4}\sin^2\theta = \frac{s^2}{2\,\Lambda^4}\frac{\sin^4\theta}{1 - \cos^2\theta}$$

In both cases (a) and (b) the modifications at $\theta = 90°$ are of equal size and maximum. They are numerically smaller than in the case of Bhabha scattering or lepton pair production (e.g. $\Lambda = 100$ GeV, $\sqrt{s} = 31$ GeV: $\delta \sim 0.4$ % at $\theta = 90°$) which is reflected in correspondingly lower experimental limits on Λ.

In summary we can say: Bhabha scattering is sensitive to QED modifications both in the time-like and space-like region of q^2. Only products of vertex and pro-pagator modifications are measured in O (q^2/Λ^2). The same statements are valid for lepton pair production, but only for $q^2 = s > 0$. The two-photon annihila-tion allows tests of QED due to couplings to neutral and charged objects. Here the modifications show up in O (q^4/Λ^4). The electron propagator is not tested[10].

3. Experimental results

Bhabha scattering, $\mu\mu$ and $\tau\tau$ pair production and two-photon annihilation have been studied by four experimental groups at PETRA (i.e. JADE, MARK J, PLUTO and TASSO). Experimental results for the s dependence of the differential cross sections are available up to $\sqrt{s}$ = 31.6 GeV (MARK J: $\sqrt{s}$ = 35 GeV) and are either published or submitted for publication[3,4,5,6]. The following figures repre-sent a selected sample and are intended to illustrate the statistical signifi-cance of the present data.

Fig. 1 shows s $d\sigma/d\Omega$ for Bhabha scattering as measured by TASSO. The solid curve is the QED prediction. The dashed curves were calculated using $\Lambda_{\pm}$ = 100 GeV. In fig. 2 the s dependence of the μ pair cross section is displayed (the entries are from all experiments). The curves are similar to those shown in fig. 1. The differential μ pair cross section s $d\sigma/d\Omega$ for TASSO and JADE is shown in fig. 3. The solid curve corresponds to the QED expectation. In fig. 4 the s dependence of $\sigma(e^+e^- \to \tau\tau)$ as measured by MARK J, PLUTO and TASSO is compared with QED (solid curve). Fig. 5 shows s $d\sigma/d\Omega$ for $e^+e^- \to \gamma\gamma$ (JADE and PLUTO). The solid curve is the QED prediction, the dashed curves indicate the deviations for a Λ value of 40 GeV. Notice the effect of the parameterisation: $\delta_{\gamma\gamma} \sim (s^2/2\Lambda^4) \cdot \sin^2\theta$ (E^*-graph in section 2). A much smaller value of Λ produces here the same ef-fect as e.g. $\Lambda_{\pm}$ = 100 GeV in the case of μ pair production ($\delta_{\mu\mu} \sim 2 \cdot s/\Lambda_{\pm}^2$ in fig. 2). All data points in fig. 1 to 5 are radiatively corrected.

The fitted values for δ are used to derive lower bounds on Λ in reactions (1) to (3) which are summarized in table 2. These bounds are defined as lower limits for Λ on the 95 % confidence level and can be asymmetric. Obviously, small de-viations from the QED prediction ($\delta = \Delta\sigma/\sigma_{QED}$; $\Delta\sigma = \sigma_{meas} - \sigma_{QED}$) correspond to large central values of Λ and also to large bounds, if the error of $\Delta\sigma/\sigma$ is small. The cut-off parameters depend on the actual parameterisation used which is different in different experiments (column 2 of table 2). Hence care must be taken in a comparison. Moreover, the quoted values for the bounds have to be judged with caution, if they are both very big (e.g. $\Lambda_+ > 200$ GeV). This is because of systematic uncertainties, polarisation effects and/or electro-weak interference.

To give an quantitative example, let us consider μ pair production. Here $\delta \simeq 2 \, s/\Lambda^2 \rightsquigarrow \Lambda \simeq (\frac{2 \, s}{\Delta\sigma/\sigma})^{1/2}$ independent of the scattering angle θ.

Lower bounds on Λ are obtained qualitatively by varying $\Delta\sigma/\sigma$ within $\sim$ 2 standard deviations. They are large if $\sqrt{s}$ is big and if $\Delta\sigma/\sigma$ and its error are small. Example given: $\Delta\sigma/\sigma = (5 \pm 5)$ %, $s = 1000$ GeV$^2 \rightsquigarrow \Lambda = 200$ GeV (central value) and $\Lambda_+ = 115$ GeV, $\Lambda_- = 200$ GeV (95 % c.l.). Consequently both lower bounds $\Lambda_\pm$ can never simultaneously exceed 140 GeV at this energy if the systematic error of $\Delta\sigma/\sigma$ is of the order of 5 %. Such uncertainties, however, are hard to exclude due to the usual problems with normalization (luminosity), background and experimental acceptance.

Another reason for a systematic effect in the determination of Λ and its bounds would be a longitudinal beam polarisation $p_\shortparallel$. The μ pair cross section is proportional to $(1 - p_\shortparallel^2)^{(12)}$, hence a 22 % polarisation results in $\Delta\sigma/\sigma = 5$ %. Again both bounds cannot exceed 140 GeV!

Finally, electro-weak effects introduce already sizeable deviations from the QED predictions. These effects are calculable in the framework of the Weinberg-Salam model (see next section) and are of the order of 2 to 5 % at $\sqrt{s} = 31$ GeV, depending on the Weinberg-angle θ_W. If they are not properly accounted for, the quoted bounds on Λ will be affected.

Another example is Bhabha scattering in which the deviations from QED are angle dependent. Figs. 6a) and b) show the differential cross section (PLUTO) at 30 GeV and 31.6 GeV. In figs. 7a) to d) the ratio $1+\delta = (d\sigma/d\Omega)/(d\sigma/d\Omega)_{QED}$ is plotted versus $\cos\theta$, together with the expectations for different values of $\Lambda_\pm$, $p_\shortparallel$ and $\sin^2\theta_W$. Clearly, cut-off parameters can only be extracted if assumptions on $p_\shortparallel$ and $\sin^2\theta_W$ are made.

Taking the average of the $\Lambda_\pm$ values for Bhabha scattering and μ pair production, as listed in table 2, and assuming that most of the unknown systematic uncertainties have been taken into account, we can estimate Λ conservatively to be ≥ 100 GeV. This means that QED has been tested successfully to distances $\lambda < 2 \cdot 10^{-16}$ cm.

4. Electro-weak interference

At the highest PETRA energies of $\sqrt{s} \sim 35$ GeV the weak interaction is expected to give sizeable effects due to electro-weak interference in the diagrams:

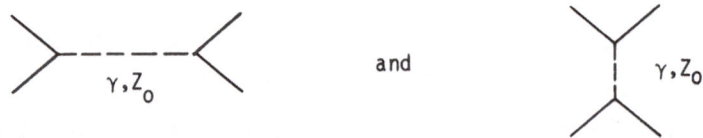

and

In unified field theories the effective Langrangian for any final state $f\bar{f}$ is conveniently written as:

$$\mathcal{L} = -e \, A^\mu \, \bar{f} \, \gamma_\mu \, f - M_z \left(\frac{G}{\sqrt{2}}\right)^{1/2} z^\mu \, \bar{f} \gamma_\mu \frac{1}{\sqrt{2}} (v - a \, \gamma_5) \, f \qquad (14)$$

where the fields A^μ, z^μ describe the photon and the vector boson Z_0. M_z is the Z_0 mass and v, a are the vector-, axial-vector coupling constants.

In the Weinberg-Salam model[13] M_z, v and a are expressed via a single parameter, θ_W (the Weinberg angle), as:

$$M_z = 74.6/\sin(2\theta_W); \qquad v = 4 \sin^2\theta_W - 1; \qquad a = -1$$

Experimentally e.g. in ν physics: $\sin^2\theta_W \approx 0.23$, hence $v \ll 1$. The above Lagrange density leads to the following differential cross section for Bhabha scattering and μ pair production[14]:

a) $e \, e \rightarrow e \, e$:

$$\frac{4s}{\alpha^2} \frac{d\sigma}{d\Omega} = \left[\left\{\frac{3 + x^2}{1 - x}\right\}^2 + 2 \frac{3 + x^2}{(1 - x)^2} \{ (3 + x) \, Q - x \, (1 - x) \, R \} v^2 \right.$$

$$- \frac{2}{1 - x} \{ (7 + 4x + x^2) \, Q + (1 + 3x^2) \, R \} a^2$$

$$(15)$$

$$+ \frac{1}{2} \{ \frac{16}{(1 - x)^2} \, Q^2 + (1 - x)^2 \, R^2 \} (v^2 - a^2)^2$$

$$\left. + \frac{1}{2} (1 + x)^2 \{ (\frac{2}{1 - x} \, Q - R) \}^2 \, (v^4 + 6v^2 \, a^2 + a^4) \right].$$

$$\cdot \, (1 + \delta_{rad} + \delta_{had})$$

b) $e e \rightarrow \mu \mu$:

$$\frac{4s}{\alpha^2} \frac{d\sigma}{d\Omega} = \left[(1 + x^2) \{1 + 2v^2 R + (v^2 + a^2)^2 R^2\} \right.$$
$$\left. + 4x \{a^2 R + 2v^2 a^2 R^2\} \right] (1 + \delta_{rad} + \delta_{had})$$

(16)

with $Q = \rho M_Z^2 \dfrac{q^2}{q^2 - M_Z^2} \rightarrow -\rho q^2$ $\quad\quad (M_Z^2 >> q^2)$

$R = \rho M_Z^2 \dfrac{s}{s - M_Z^2} \rightarrow -\rho s$ $\quad\quad (M_Z^2 >> s)$

$\rho = \dfrac{G}{8\sqrt{2}\,\pi\alpha} = 4.49 \cdot 10^{-5}$ (GeV^{-2}); $x = \cos\theta$; G = universal Fermi coupling constant

The width of the Z_0 boson is neglected in the propagators Q and R. In eqs. (15) and (16) the first term corresponds to the pure QED contribution and the terms linear in Q and R describe the electro-weak interference. For practical applications radiative corrections in the interference terms are neglected.

By inspection of eqs. (15) and (16) we can state:

i) The differential cross sections depend on M_Z, v^2, a^2, but not on the sign of the coupling constants which show up only in parity violating polarization terms.

ii) The contribution of the direct Z_0 term is proportional to s ($M_Z^2 >> s, q^2$) as expected for a point-like weak coupling.

iii) Bhabha scattering is more sensitive to v^2 than μ pair production.

iv) The interference term in $ee \rightarrow \mu\mu$ produces a forward-backward asymmetry A which is sensitive to a^2:

$$A = \frac{F - B}{F + B} \approx \frac{3}{4} \frac{2a^2 R}{1 + 2v^2 R} < 0 \quad\quad (in\ O(R))$$

v) The total μ pair cross section can be related in a simple way to cut-off parameters $\Lambda_\pm$. Integration of eq. (16) yields ($M_Z^2 >> s$):

$$\sigma_{\mu\mu} = \sigma_{QED} \left(1 - 2v^2 \rho s + (v^2 + a^2)^2 \rho^2 s^2\right)$$

$$= \sigma_{QED} \left(1 \mp \frac{s}{s - \Lambda_\pm^2}\right)^2 \sim \sigma_{QED} \left(1 \pm \frac{2s}{\Lambda_\pm^2}\right)$$

$$\pm\left(\frac{1}{\Lambda_\pm^2}\right) = -\rho v^2 + \frac{1}{2} \rho^2 s (v^2 + a^2)^2 \qquad (17)$$

Note that the correction changes sign with increasing s.

So far these statements are model independent. If we assume the validity of the Weinberg-Salam model, we can estimate how big the electro-weak effects are at present PETRA energies (s ~ 1000 GeV^2). With sin$^2\theta$ = 0.23 we have:

M_Z = 88.6 GeV

v^2 = 0.0064

a^2 = 1

Inserting these numbers into the cross sections (eqs. (15) and (16)) we obtain:

- The corrections to Bhabha scattering due to electro-weak interference are ~ 2 % (see fig. 1 and fig. 7d), thus at the limit of detectability.

- The μ pair asymmetry is ~ -8 %. Radiative corrections ~ α^3 and the limited angular acceptance reduce this value to ~ -4 % which is hard to measure with present statistics.

- The cut-off parameter Λ_+ (eq. (17)) corresponding to sin$^2\theta_w$ = 0.23 is: Λ_+ = 1170 GeV, far above the quoted bounds in table 2.

So, if we believe in unified models with sin$^2\theta_w$ = 0.23, we can presently neglect the weak interaction for the total cross sections even at highest PETRA energies, except for a small correction in Bhabha scattering at large angles. Nevertheless, it is legitimate to question the accepted value for sin$^2\theta_w$, or to question the Weinberg-Salam model altogether. In order to study which experimental limits can be placed on these quantities, we change our view point somewhat and assume the validity of QED to avoid additional unknown parameters. Specifically, the simultaneous analysis of Bhabha scattering and μ pair production at highest PETRA energies is directed to the following problems:

- determination of upper bounds on $\sin^2\theta_w$
- fit of v^2 and a^2 ($M_Z^2 \gg s$, q^2) in a model independent way
- search for new Z_o poles at low masses[15]

Through the use of eq. (17) for lepton-pair production all PETRA groups quote upper limits on $\sin^2\theta_w$ ranging from 0.49 (JADE) to 0.70 (PLUTO) (95 % c.l.). Preliminary results on the coupling constants v^2 and a^2 have been reported by JADE[16]. The fitted values are:

$$a^2 = 0.9 \pm 1.4 \quad (90 \text{ % c.l.})$$

$$v^2 = 0.1 \pm 0.6 \quad (90 \text{ % c.l.})$$

Data taking continues at PETRA up to energies of $\sqrt{s}$ = 37 GeV and more significant results are expected in the near future.

5. Summary

The present PETRA experiments on Bhabha scattering, lepton-pair production and two-photon annihilation are consistent with exact QED up to $|q^2| \sim 1000$ GeV^2.

Lower bounds on the cut-off parameters Λ exceed 100 GeV which is equivalent to tests of QED down to distances < $2 \cdot 10^{-16}$ cm.

Electro-weak interference effects become sizeable at the highest PETRA energies.

Further tests of QED up to LEP energies are desirable and not impossible, providing weak effects are properly isolated. The two-photon annihilation process is particularly well suited for such studies, since $Z_o\gamma$ interference is absent here in first order perturbation theory.

Acknowledgement

I am indebted to my colleagues at DESY (in particular at PLUTO) who helped me in preparing this talk. I am very grateful to Dr. F. Gutbrod for many helpful discussions. I want to thank the DESY directorate for their hospitality extended to me during my stay in Hamburg. I am grateful to Prof. G. Gräff for the invitation to the Mainz Symposium on the present status of QED.

References

(1) R. Hofstadter, Proc. 1975 Int. Symposium on Lepton and Photon Inter-
        actions at High Energies, Stanford, USA, 1975, p. 869

(2) J.E. Augustin et al., Phys. Rev. Lett. 34 (1975) 233
        T. Himel et al., Phys. Rev. Lett. 41 (1978) 449
        B.L. Beron et al., Phys. Rev. D17 (1978) 2187 and 2839

(3) W. Bartel et al. (JADE Coll.), DESY-rep. 80/14 (1980), to be published, and
        P. Dittmann, private communication

(4) D.P. Barber et al. (MARK J Coll.), Phys. Rev. Lett. 43 (1979) 1915 and
        J. Bron, private communication

(5) Ch. Berger et al. (PLUTO Coll.), DESY-rep. 80/01 (1980) and 80/35 (1980),
        to be published

(6) R. Brandelik et al. (TASSO Coll.), DESY-rep. 80/33 (1980), to be
        published

(7) Ch. Berger et al. (PLUTO Coll.), DESY-rep. 80/34 (1980), to be published

(8) J. Takahashi, Nuov. Cim. 6 (1957) 370

(9) F.E. Low, Phys. Rev. 110 (1958) 974

(10) N.M. Kroll, Nuov. Cim. XLV (1966) 65
        K. Ringhofer, H. Salecker, contributed paper 109 to the 1975 Int. Sympo-
        sium on Lepton and Photon Interactions at High Energies, Stanford,
        USA, 1975

(11) A. Litke, Harvard University Ph. D. Thesis (unpublished) 1970

(12) K. Koller, T.F. Walsh, P.M. Zerwas, DESY int. rep. T-79/01 (1979)

(13) S.L. Glashow, Nucl. Phys. 22 (1961) 579
        S. Weinberg, Phys. Rev. Lett. 19 (1967) 1264
        A. Salam, Proc. 8th Nobel Symposium, N. Svartholm, editor, Wiley N.Y.
        (1968)

(14) R. Budny, Phys. Lett. 55B (1975) 227

(15) E.H. de Groot, D. Schildknecht, G.J. Gounaris, U. Bielefeld preprint
        BI-TP 79/37 (1979)

(16) R. Marshall, talk presented at Rencontre de Moriond, 1980

Table 1   Existing and planned ee storage rings

Ring	Start of operation			Beam energy (GeV)
Ada	Frascati	1960	e^+e^-	0.25
Princeton-Stanford	Stanford	1962	e^-e^-	0.55
ACO	Orsay	1966	e^+e^-	0.2 - 0.55
VEPP-2	Novosibirsk	1966	e^+e^-	0.2 - 0.55
ADONE	Frascati	1969	e^+e^-	0.7 - 1.55
BYPASS	Cambridge (USA)	1971	e^+e^-	~1.5 - 3.5
SPEAR	Stanford	1972	e^+e^-	1.2 - 4.2
DORIS	Hamburg	1974	e^+e^-	~1.5 - 5.1
VEPP-2M	Novosibirsk	1975	e^+e^-	0.2 - 0.67
DCI	Orsay	1976	e^+e^-	0.5 - 1.7
VEPP-4	Novosibirsk	1978	e^+e^-	similar to CESR
PETRA	Hamburg	1978	e^+e^-	5 - 19
CESR	Cornell	1979	e^+e^-	3 - 8
PEP	Stanford	1980	e^+e^-	5 - 18
LEP	Geneva	1986?	e^+e^-	22 - 130

Table 2   PETRA results on QED cut-off parameters

reaction	group	definition of $\Lambda_\pm$	Λ_+ (GeV)	Λ_- (GeV)
$e^+e^- \to e^+e^-$	JADE	$F_s = 1 \pm q^2/(q^2 - \Lambda_\pm^2)$ $F_t = 1 \pm s/(s - \Lambda_\pm^2)$	> 104	> 87
	MARK J	$F_s = 1 \mp q^2/(q^2 - \Lambda_\pm^2)$ $F_t = 1 \mp s/(s - \Lambda_\pm^2)$	> 74	> 95
	PLUTO	$F_s = 1 \pm q^2/\Lambda_\pm^2$ $F_t = 1 \pm s/\Lambda_\pm^2$	> 80	> 234
	TASSO	$F_s = 1 \mp q^2/(q^2 - \Lambda_\pm^2)$ $F_t = 1 \mp s/(s - \Lambda_\pm^2)$	> 112	> 139
$e^+e^- \to \mu^+\mu^-$	MARK J	$F_t = 1 \mp s/(s - \Lambda_\pm^2)$	> 129	> 137
	PLUTO	$F_t = 1 \pm s/\Lambda_\pm^2$	> 87	> 99
	TASSO	$F_t = 1 \mp s/(s - \Lambda_\pm^2)$	> 80	> 118
$e^+e^- \to \tau^+\tau^-$	MARK J	$F_t = 1 \mp s/(s - \Lambda_\pm^2)$	> 82	> 120
	PLUTO	$F_t = 1 \pm s/\Lambda_\pm^2$	> 74	> 65
	TASSO	$F_t = 1 \mp s/(s - \Lambda_\pm^2)$	> 73	> 82
$e^+e^- \to \gamma\gamma$	PLUTO	"sea-gull" $F(q^2) = 1 \pm q^4/\Lambda_\pm^4$	> 46	> 36
	JADE	heavy electron: $\delta_\Lambda = \pm s^2 \sin^2\theta/(2\,\Lambda_\pm^4)$	> 45	> 38
	PLUTO	$\delta_\Lambda = +s^2 \sin^2\theta/(2\,\Lambda_\pm^4)$	> 46	-
	TASSO	$\delta_\Lambda = \mp s^2 \sin^2\theta/(2\,\Lambda_\pm^4)$	> 34	> 42

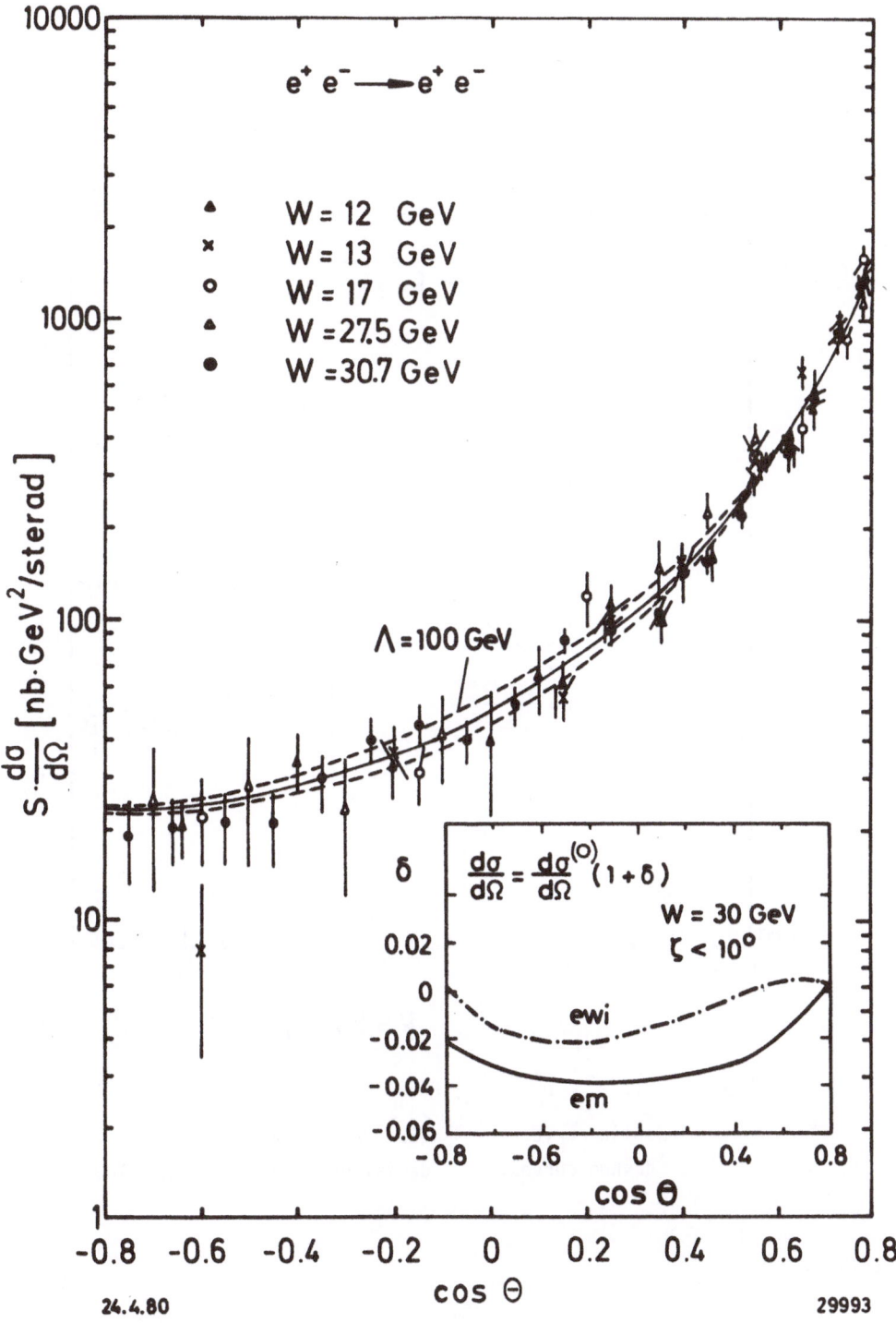

Fig. 1 $s \cdot \frac{d\sigma}{d\Omega} (e^+e^- \rightarrow e^+e^-)$ as measured by TASSO

solid curve: QED prediction
dashed curves: deviation from QED, if $\Lambda = 100$ GeV
insert: radiative (em) and electro-weak (ewi) corrections

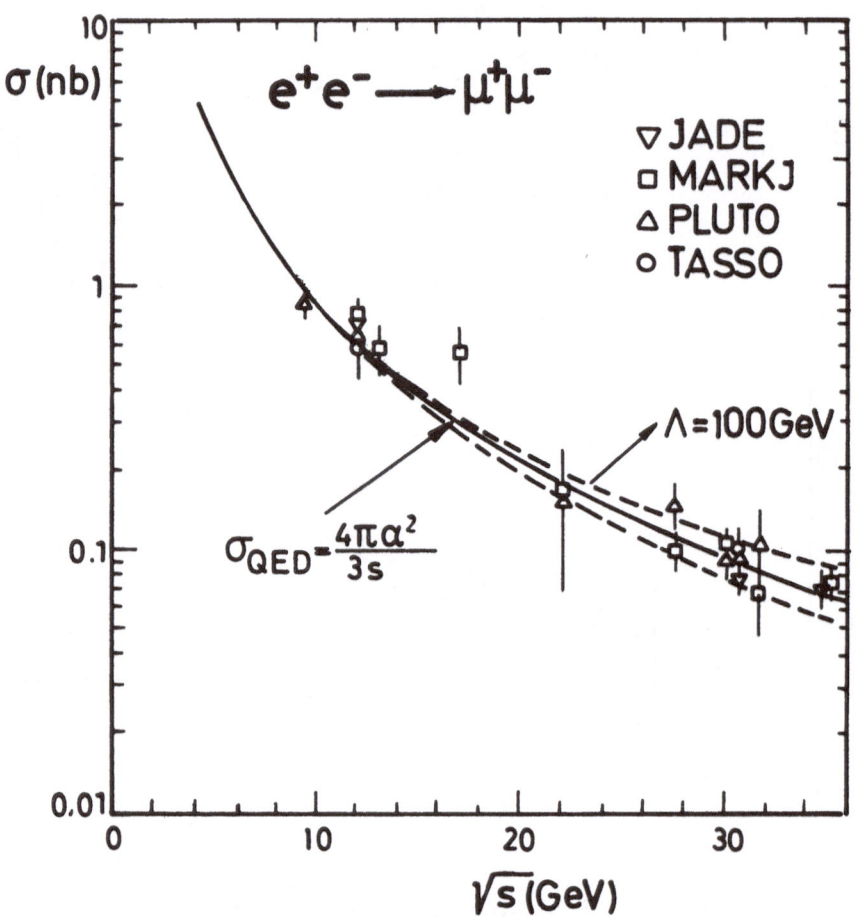

Fig. 2 $\sigma(e^+e^- \rightarrow \mu^+\mu^-)$ all PETRA groups

solid curve: QED prediction
dashed curves: deviation from QED, if Λ = 100 GeV

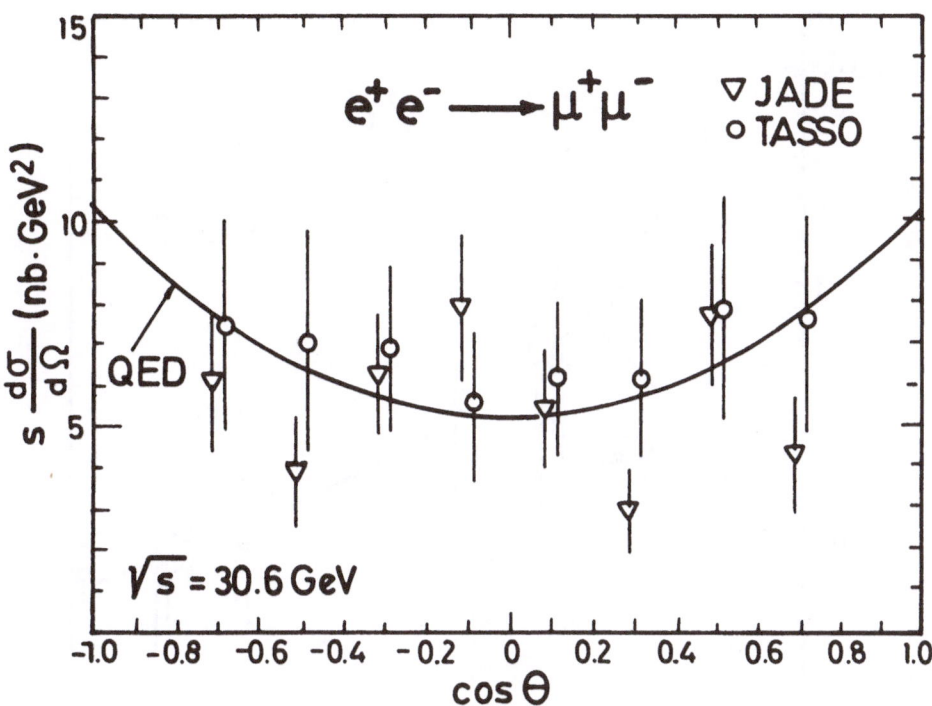

Fig. 3 $s \cdot \frac{d\sigma}{d\Omega}$ ($e^+e^- \rightarrow \mu^+\mu^-$) as measured by JADE, TASSO at $\sqrt{s}$ = 30.6 GeV

solid curve: QED prediction

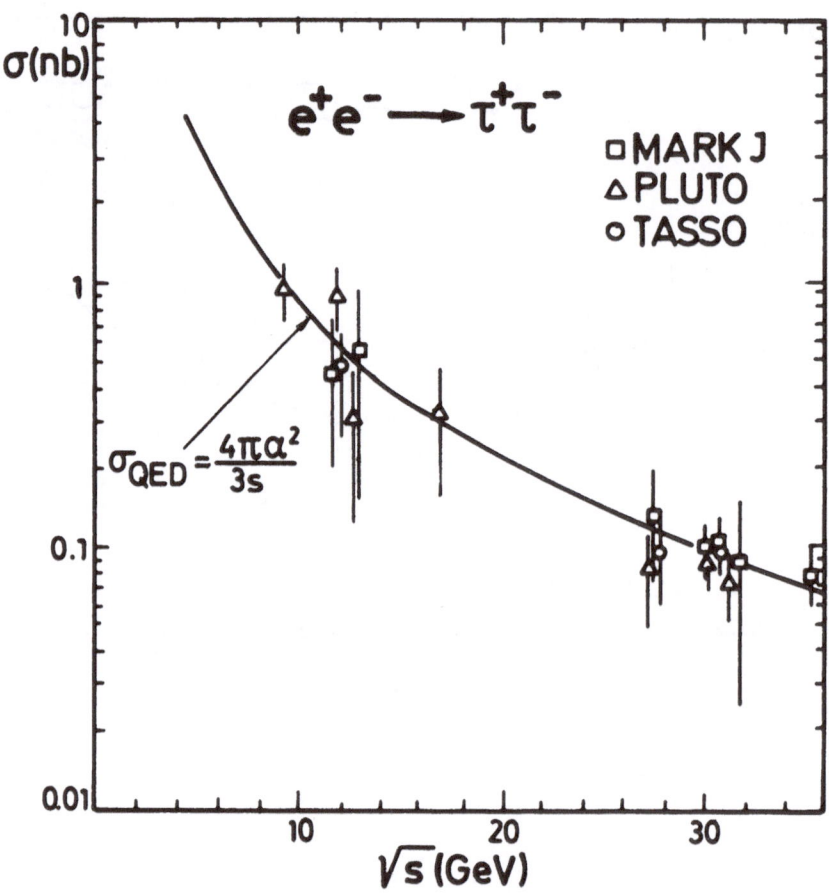

Fig. 4 $\sigma(e^+e^- \rightarrow \tau^+\tau^-)$ as measured by MARK J, PLUTO, TASSO

solid curve: QED prediction

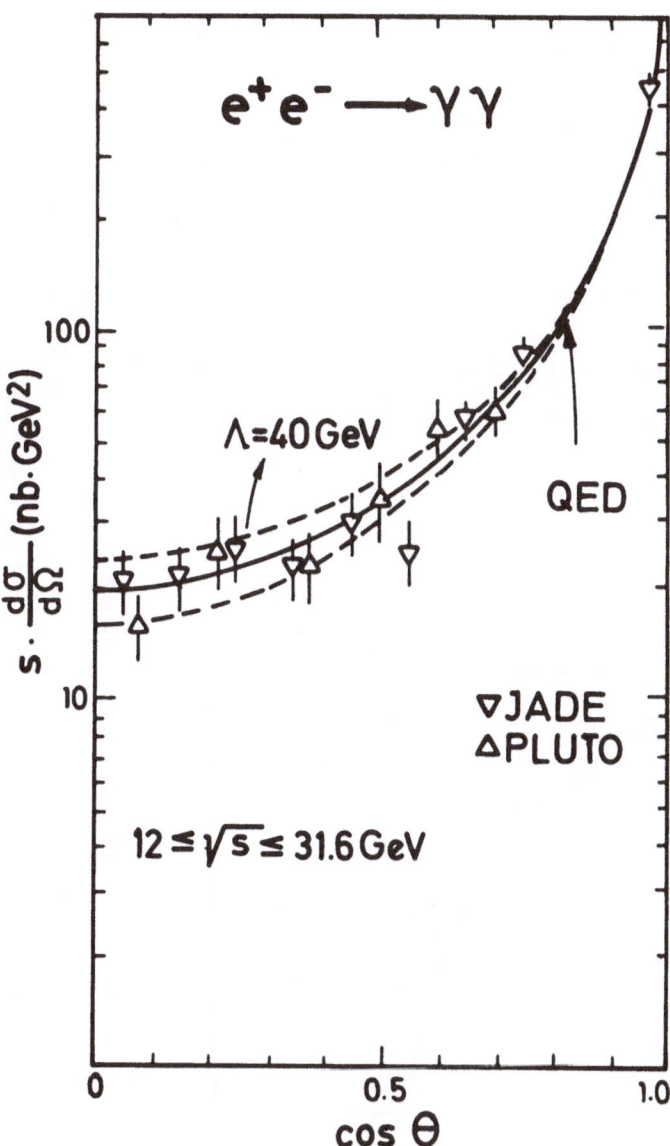

Fig. 5 $s \cdot \frac{d\sigma}{d\Omega}$ ($e^+e^- \rightarrow \gamma\gamma$) as measured by JADE, PLUTO ($12 \leq \sqrt{s} \leq 31.6$ GeV)

solid curve: QED prediction

dashed curves: deviation from QED, if Λ = 40 GeV

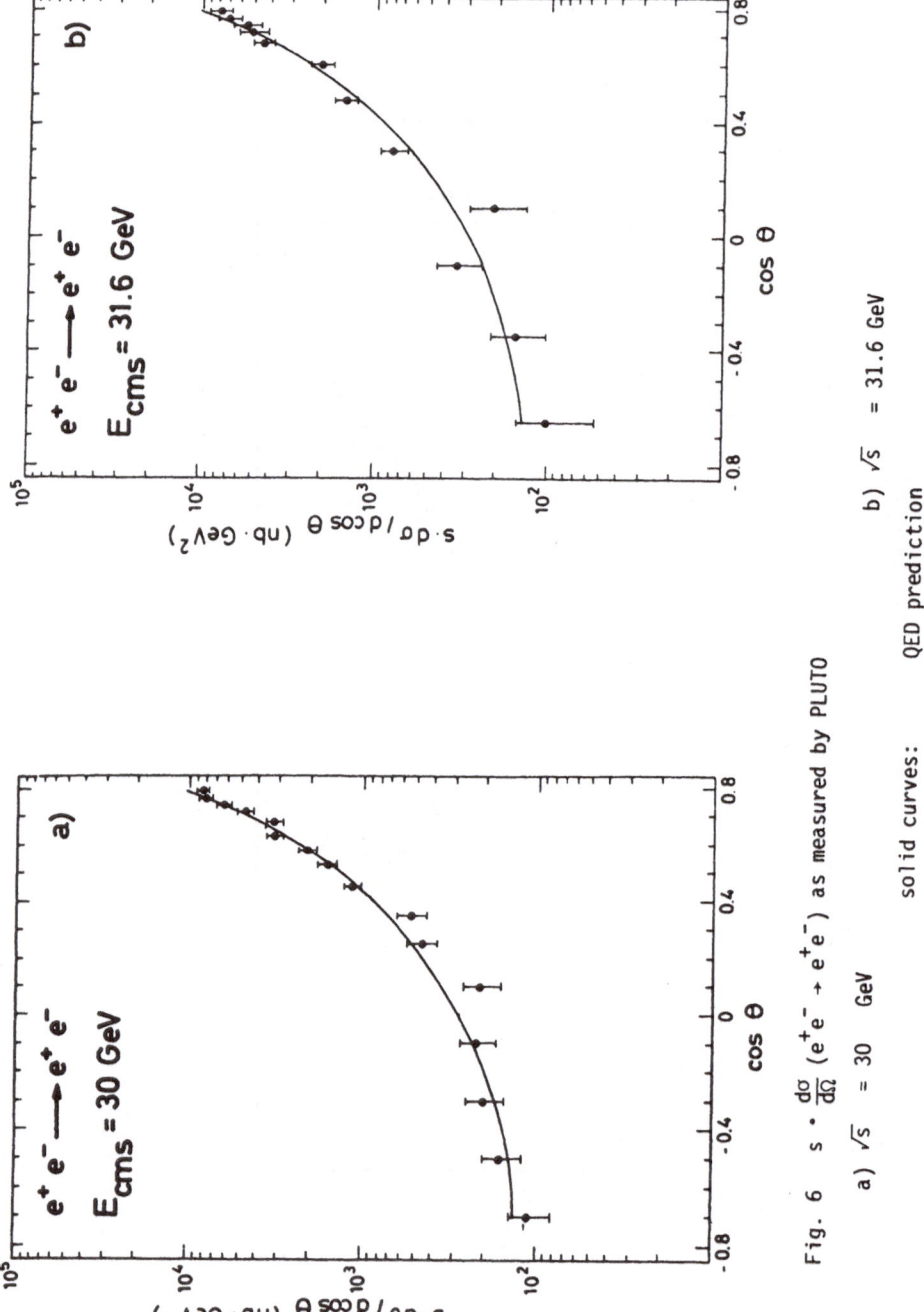

Fig. 6 $s \cdot \frac{d\sigma}{d\Omega} (e^+e^- \to e^+e^-)$ as measured by PLUTO

a) $\sqrt{s}$ = 30   GeV

b) $\sqrt{s}$ = 31.6 GeV

solid curves: QED prediction

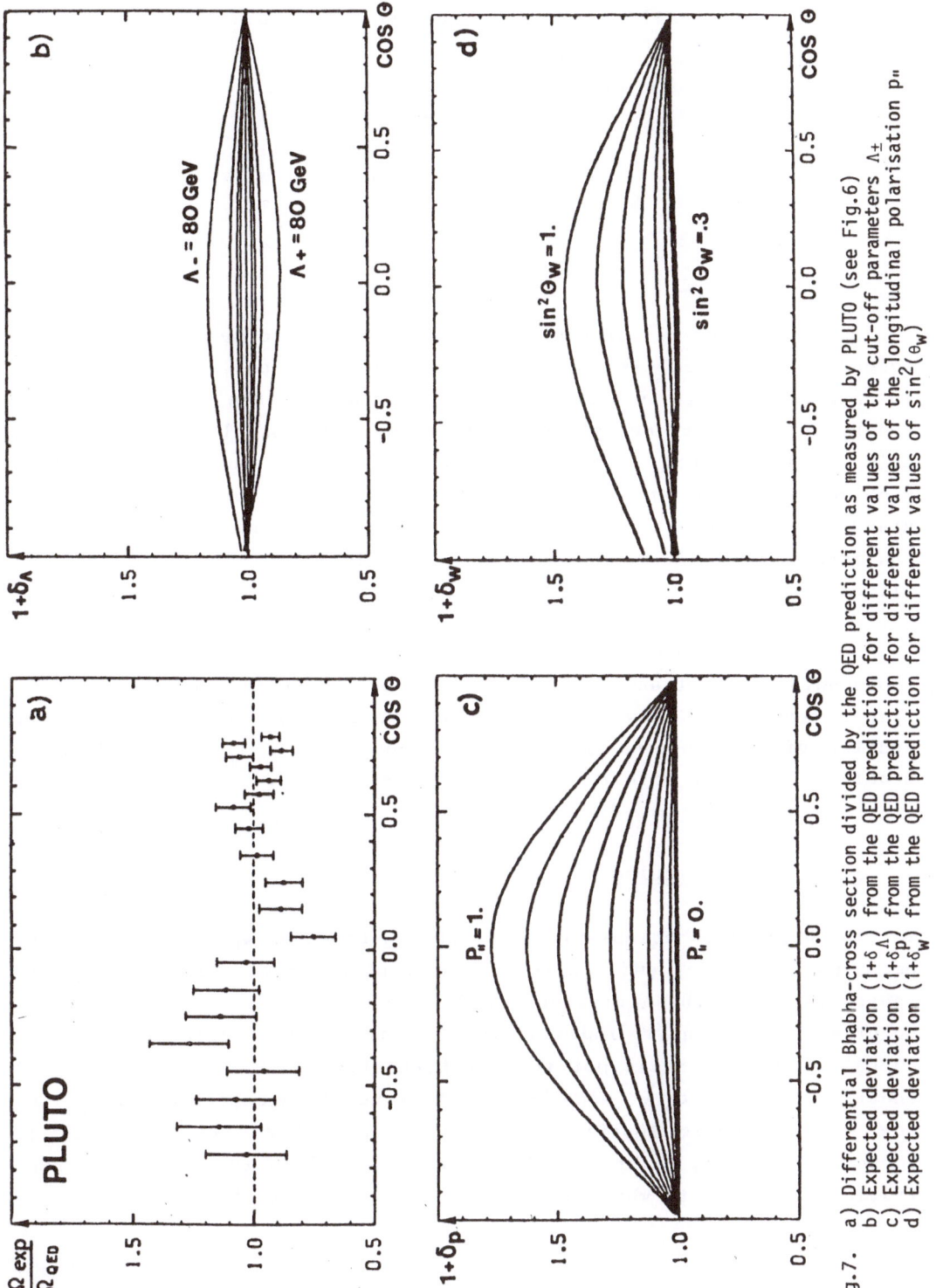

Fig.7. a) Differential Bhabha-cross section divided by the QED prediction as measured by PLUTO (see Fig.6)
b) Expected deviation $(1+\delta_{\Lambda})$ from the QED prediction for different values of the cut-off parameters $\Lambda_{\pm}$
c) Expected deviation $(1+\delta_{p})$ from the QED prediction for different values of the longitudinal polarisation $p_{\shortparallel}$
d) Expected deviation $(1+\delta_{w})$ from the QED prediction for different values of $\sin^{2}(\theta_{w})$

SOME BASIC PROBLEMS OF QUANTUM ELECTRODYNAMICS

O. Steinmann
Fakultät für Physik
Universität Bielefeld
4800 Bielefeld

QED (=quantum electrodynamics) is often said to be one of the most successful theories, if not "the" most successful one, that we possess in physics. That it is highly successful cannot be denied. That it is a theory is, however, not yet established beyond possible doubt. In this talk I will report, very cursorily, on the present state of this problem. The question is whether the computational rules of QED, which stand up so well to all practical tests, can be founded logically in a consistent, exactly formulizable, theory. In other words: does QED at all have the right to work as well as it does?

I wish to touch three problems which are of interest in this fundamental context and which at present are being investigated more or less intensively. They are: first and most important, the existence problem of QED, second the infrared problem, and third, a class of problems connected with the gauge structure of QED.

1. Existence problem.

The well-known calculations of QED are based on approximations like perturbation theory and others. The evident question arises, <u>what</u> gets approximated by these procedures, i.e. whether there exists an exact solution of QED, quite independently of its explicit calculability. It turns out that this is not the first question to be asked. The first question is: what do we mean when we talk about an exact solution, i.e. how are the basic equations of the theory, which this solution is supposed to solve, to be formulated correctly? This is an entirely non-trivial question, to which we have as yet no answer.

A first attempt at such a formulation will roughly look like this: QED is a field theory, whose fundamental fields are the 4-potential A_μ and the Dirac spinors ψ, $\bar{\psi} = \psi*\gamma^0$. The field strenghts $F_{\mu\nu}$ are related to A_μ by the familiar formulae

$$F_{\mu\nu}(x) = \delta_\mu A_\nu(x) - \delta_\nu A_\mu(x) . \tag{1}$$

The fields must solve the equations of motion

$$\delta_\beta F^{\alpha\beta}(x) = - A^\alpha(x) + \delta^\alpha \delta_\beta A^\beta(x) = j^\alpha(x) \tag{2}$$

$$(i\not{\delta}-m) \psi(x) = -e \not{A}(x)\psi(x) , \tag{3}$$

where j^α is the current density

$$j^\alpha(x) = e \bar{\psi}(x) \gamma^\alpha \psi(x) . \tag{4}$$

x is the 4-vector $(x^0=ct, x)$. Note that we shall use throughout the Heisenberg picture as the most appropriate picture in a relativistic situation.

A solution of the theory is a solution of these equations of motion, possibly satisfying some physically motivated subsidiary conditions. This formulation is satisfactory as long as we are dealing with c-number valued fields, i.e. in a first-quantized theory, in which the electromagnetic field is treated classically and the electron wave-mechanically. But in QED the fields are operator valued, and this creates no end of problems. This is most easily seen as follows. In a quantum field theory we demand besides the field equations also the validity of canonical commutation relations. To each field f_σ (f_σ denotes henceforth an arbitrary component of A_μ, ψ, or $\bar{\psi}$) we associate, according to certain fixed rules, a conjugate field π_σ , and we demand relations which are typically of the form

$$[f_\sigma(x ,0) , \pi_\tau(y ,0)]_\pm = i \hbar \delta_{\sigma\tau} \delta^3(x-y) . \tag{5}$$

Actually the correct commutators of QED look somewhat more complicated, depending on the gauge chosen. But this is irrelevant for our present purpose. What matters is that the right-hand side of (5) is at least as singular as a δ-"function". But δ isn't a function, it is a distribution. The same is then true for the left-hand side of (5): our fields cannot be functions of x , they are distributions. This means that the value of f_σ at a point x is <u>not</u> defined. Mathematically meaningful are only the averages $\int d^3x\, f_\sigma(x,x^0)\,\varphi(x)$, or even $\int d^4x\, f_\sigma(x)\,\varphi(x)$, over sufficiently well-behaved "test-functions" φ . Typically the admissible φ are differentiable a certain number of times and decrease sufficiently fast for $x\to\infty$. Now, distributions have the unpleasant property that they can in general not be multiplied one with the other. E.g., the product $[\delta(x)]^2$ of a δ-function with itself is not defined. But exactly such products appear in the right-hand sides of the field equations (2) and (3), and these right-hand sides are therefore at first undefined. If we calculate nevertheless with these meaningless equations, we are punished by the appearance of the well-known ultra-violet divergences.

In order to arrive at an exact formulation of QED it is thus imperative to find a better definition of the right-hand sides of the field equations. Let us discuss this briefly for the example of the current $j^\alpha(x)$. Let us first try to set down the properties that an operator valued distribution $j^\alpha(x)$ must necessarily possess in order to be acceptable as a candidate for the current operator.

a) The current j^α and the field $F^{\mu\nu}$ are observables, hence j^α must be what is called a local field, i.e. we must have

$$[j^\alpha(x)\ ,\ j^\beta(y)] = [j^\alpha(x)\ ,\ F^{\mu\nu}(y)] = 0 \tag{6}$$

for space-like separations $x-y$. This is so because measurements carried out in relatively space-like regions cannot disturb one another, so that $j^\alpha(x)$ on the one hand and $j^\beta(y)$ or $F^{\mu\nu}(y)$ on the other hand are measurable simultaneously.

b) j^α must transform as 4-vector under Lorentz transformations.

c) The continuity equation

$$\delta_\alpha\, j^\alpha(x) = 0$$

must be satisfied.

d) The component j^0 must represent the charge density, i.e. the charge operator Q is given by

$$Q = \int d^3x\, j^0(x,t)\ , \tag{8}$$

which expression is time independent by virtue of (7) (charge conservation!). Q being the charge operator is expressed through the requirement

$$[Q, A_\mu(x)] = 0 ,$$

$$[Q, \psi(x)] = e \ \psi(x) \quad , \quad [Q , \bar\psi(x)] = -e \ \bar\psi(x)$$

$$(9)$$

where e is the charge of the electron.

Two questions arise. First, do there at all exist current operators with all these properties, as operators in the same state space in which the fields A_μ, ψ, $\bar\psi$ act? Second, if such j^α exist , are they determined uniquely by our requirements? If not, which additional conditions are needed to remove the ambiguity?

In <u>perturbation theory</u> we know the answers to these questions. The current j^α is defined through replacing the naive product ansatz (4) by what is called a "normal product" :

$$j^\alpha(x) = e \ N(\bar\psi(x)\gamma^\alpha\psi(x)).$$

$$(10)$$

There are various equivalent definitions of this normal product. All of them are too complicated to be explained here in the short time available. Roughly, they are just suitable versions of the well-known renormalization prescriptions. These prescriptions can be formulated in a mathematically completely rigorous way (within the context of perturbation theory!), so that at no place divergent expressions enter, not even in intermediary steps, and so that no preliminary regularization is necessary. Uniqueness of the definition is guaranteed by certain regularity conditions, either for the local behaviour in x-space ($j^\alpha(x)$ shall show only the weakest possible local singularities compatible with the other requirements), or for the asymptotic behaviour in p-space (minimal increase at high energies). In the literature these requirements are often used tacitly, without being mentioned explicitly by the authors. It turns out that they are equivalent with the requirement of renormalizability. This is, however, no compelling reason for their adoption. We know of no physical principle which would prevent nature from using so-called non-renormalizable theories. Demanding renormalizability is merely a question of expediency, unless one wants to appeal to the metaphysical principle of the "economy of nature". It makes sense to investigate at

first the simplest among a bunch of possible theories, and if it works
as well as QED does one will gladly dispense with the study of
complicated alternatives.

The various definitions of a normal product are all more or less
specifically based on the special structures met with in perturbation
theory. It is not known whether any of these definitions can be generalized in such a way that it engenders field equations possessing <u>exact</u>
solutions. I.e. the <u>existence problem of QED is not yet solved</u>, and the
unicity problem is therefore not yet acute.

Presumably, the existence problem could be solved if the perturbation expansion converged. Remember that the perturbative procedure
consists in expanding all the relevant quantities of the theory (fields,
cross section, or whatever) in power series in the coupling constant e,
e.g.

$$f_\sigma(x) = \sum_{n=0}^{\infty} e^n f_{\sigma,n}(x) ,\qquad (11)$$

and then prescribing methods for the determination of the expansion
coefficients $f_{\sigma,n}$, etc. As mentioned above, we by now possess completely rigorous methods for the calculation of these coefficients.
However, these methods give us no information whatever on the conver-
gence of the series as a whole. If it converged and if the sum could be
shown to possess all the right properties, then this would constitute a
solution of the existence problem. But it is not known whether the
series converges. There exist arguments of various degrees of plausi-
bility pointing to divergence, and divergence seems to be the prevalent
expectation among the experts. This has, by the way, consequences for
the comparison between theory and experiment. If perturbation theory
diverges, then the computation of graphs of higher and higher order
will not necessarily lead to more and more accurate results. It could
be that at some point a limit of accuracy is reached beyond which
perturbation theory ceases to be a valid approximation.

2. Infrared problems.

From the purely field theoretical point of view there are no infrared problems in QED: as long as we are satisfied with studying the
fields and their properties, we find no infrared divergences. They
enter only if we introduce particles into the theory. But this we are
obviously forced to do in order to establish contact with reality, for
in the experiment we see electrons, not electron fields. The origin of
the difficulties and the means for their removal are quite well
understood. Since we have no rigorous QED, this understanding is of
course based on approximations, in particular on a partially summed
version of perturbation theory and on semiclassical methods. The experts
are confident, however, that the insights gained in this way are also
roughly valid for the exact theory.

The point is that electrons (more generally: all charged particles)
are "particles" in a somewhat more complicated sense than the sense
underlying the usual considerations. Naively, we (i.e. we oversophis-
ticated field theoreticians) mean by a particle an object with a
sharply defined mass: a one-particle state is a <u>normalizable</u> eigenstate
of the mass operator

$$M^2 = P_\mu P^\mu = P_o^2 - P^2 \ . \tag{12}$$

Here P_o is the energy operator, P are the momentum operators. (We
put $c = 1$.) Now, the electrons aren't particles in this sense. This
is seen most easily in the structure of the 2-point-function
$< 0|\bar\psi(x) \ \psi(y)|0 >$, or its Fourier transform

$$W(p,q) = < 0|\bar{\tilde\psi}(p)\tilde\psi(q)|0 > \ .$$

Here $|0 >$ stands for the vacuum state, and

$$\tilde\psi(q) = \int d^4y \ e^{-iqy} \ \psi(y) \ .$$

Spinor indices are suppressed, since spin is of no relevance to the in-
frared problem. Summation over a complete system $|z>$ of intermediary

states yields

$$W(p,q) = \sum_z < 0| \tilde{\psi}(p)|z > < z|\tilde{\psi}(q)|0 >$$

(13)

$$= \delta^4(p+q) \ II(q) \ .$$

The splitting-off of the factor $\delta^4(p+q)$ is simply an expression of energy-momentum conservation.

If the electron were a particle in the sense mentioned above, then II would be of the form (for $q_o > 0$, for $q_o < 0$ we have II = 0)

$$II(q) = const \cdot \delta(q^2-m^2) + \sigma(q) \ .$$

(14)

The first term on the right-hand side is the contribution of the one-electron states to the z-sum. σ is the contribution of the several-particle states and ought to behave smoothly for $q^2 \to m^2$. (m is the electron mass.) However, if we compute II in order e^2 of perturbation theory we find instead of the expected singularity

$$\delta(q^2-m^2) = - \tfrac{1}{\pi} \ Im(q^2-m^2+i\epsilon)^{-1}$$

at $q^2 = m^2$ the stronger singularity

$$Im \ \frac{log(q^2-m^2+i\epsilon)}{q^2-m^2+i\epsilon}$$

(15)

A more thoroughgoing investigation shows that this result of finite-order perturbation theory is misleading. It is possible to determine the most singular contribution to II in all orders of perturbation theory and to sum up these contributions. The result is a singularity of the form $(q^2-m^2)^{-1+\beta}$, with β = const·e^2 : a weaker singularity than shown in (14). This result is confirmed by other approximations, and we have good reason to believe that it is fairly close to reality.

The same situation concerning the strength of one-particle singularities is present also for the electron propagator, which is closely related to the function W discussed above. (We are talking here about the full, clothed, electron propagator, not the free one.) In particular, this is true for the propagators connected with the external lines of Feynman graphs. If we now compute a S-matrix element by the familiar prescriptions, multiplying external lines with (p^2-m^2) and going onto the mass shell $p^2 = m^2$, then we obtain divergent expressions in perturbation theory of finite order, vanishing expressions in partially summed up perturbation theory. Both results are equally undesirable.

The solution of the problem is as follows. Electrons aren't particles in the naive sense introduced above. An electron has quite a loose internal structure, usually visualized as a "cloud of soft photons" accompanying the charge. The splitting of a physical electron into a naked electron and a photon cloud is, however, meaningless in a strict sense. What makes sense is only the whole object. The form of the cloud (if we permit ourselves to use this convenient picture nevertheless) depends on the history of the electron, and even on the experimental arrangement used for its observation. Namely it is this arrangement which determines which photons must be considered part of the cloud, which ones are separate particles in their own right. Hence an electron state is by no means uniquely characterized by specifying its momentum and polarization. A full characterization requires detailed information on the internal structure, much more detailed information than any measurement can ever furnish. Thus the final state of a scattering experiment is not uniquely determined by the experimental setup, and we should not wonder that such experiments cannot be satisfactorily described with the help of scattering amplitudes, i.e. S-matrix elements. Such an amplitude can only give transition probabilities between well defined states. As is well known we obtain reasonable, finite, values for the scattering <u>cross sections</u>, if we add the cross sections for all the transitions which are compatible with the given experimental arrangement, hence the infrared problem disappears if only we look at the right kind of quantities.

It must be said that several methods have been proposed in the literature, which purport to save the S-matrix by defining it in a more general way than usual. But all these definitions are so complicated that they are, in my opinion, of little use either for our general understanding or for computational purposes.

3. Gauge problems

Here we want to talk about some problems, partially of a more aesthetical nature, which QED shares with the other gauge theories which are so popular at present.

As everybody knows, the field equations (2-3) are invariant under the gauge transformations

$$
\begin{aligned}
\psi(x) &\rightarrow \exp\{ie\ (x)\}\ \psi(x) \quad , \\
\bar{\psi}(x) &\rightarrow \bar{\psi}(x)\ \exp\{-ie\ (x)\} \quad , \\
A_\mu(x) &\rightarrow A_\mu(x) + \delta_\mu\ (x) \quad ,
\end{aligned}
\tag{16}
$$

where   is a real, possibly operator valued, function. Only operators which are invariant under these transformations, e.g. $F_{\mu\nu}$ or j_α, can represent observables, so that the physical content of the theory is not changed by gauge transformations. Now, the fields ψ, $\bar{\psi}$, A, have very different properties in different gauges, and these properties are not very agreeable in any gauge. E.g. in the radiation gauge the fields are neither local nor Lorentz covariant, in Gupta-Bleuler gauges they operate in a too large state space with indefinite metric, etc. It is evident that these fields have no very direct physical significance. The desire to eliminate them from the theory as far as possible is therefore natural. Attempts in this direction can roughly be divided into two categories.

Firstly, one can try to get rid of ψ and A completely and to formulate the theory exclusively in terms of the observables like $F_{\mu\nu}$ and j_α. Such observables formulations exist as general background theories and have yielded important insights, e.g. on the structure of superselection rules. It is, however, still completely unclear how to describe the dynamics of a specific model in this framework. Without ψ we cannot write down a Lagrangian or sufficiently stringent field equations, and we do not know, by what other means this dynamical information can be injected into the theory.

Secondly, and less radically, we can try to keep the bothersome fields around, but to formulate the theory in such a way that it becomes obvious at once that the important things are not the ψ and A themselves, but only their equivalence classes with respect to gauge transformations. Here there exist two promising starting points.

a) For the case of classical electrodynamics (more exactly: the firstquantized version introduced at the beginning of this talk) a suitable language for the formulation of these structures is furnished by modern differential geometry, in particular by the theory of fibre bundles. Unfortunately, this way of looking at things needs, in its current formulation, that the value of a field at a point is defined. It can therefore not be taken over easily into the quantized version of the theory. This problem remains unsolved.

b) Another very elegant formulation is possible for the so-called lattice gauge theories. Here the space-time continuum is approximated by a discrete lattice of points. This destroys Lorentz invariance, but brings many advantages in other respects. However, in order to describe the world as it is, we must be able to go over in the end to the continuum limit by letting the lattice constant (i.e. the distance between the lattice points) tend towards zero. This problem is also unsolved. It is not known whether this limit exists, and if yes, whether it can be attained in such a way that the elegance of the formulation is preserved. This is a problem on which research is very active at present, so that we can hope for essential new results at any time.

THE ANOMALOUS MAGNETIC MOMENT OF THE LEPTONS:
THEORY AND NUMERICAL METHODS

E. Borie

Institut für Theoretische Kernphysik,
Universität Karlsruhe

J. Calmet

Institut für Informatik I, Universität Karlsruhe
and CNRS, Marseille

The anomalous magnetic moment of a lepton is interesting because it would be zero in the absence of interaction with the quantized radiation field (emission and absorption of virtual photons). As a result of this interaction the leptons acquire effective form factors. This can be seen by considering the most general lepton-photon coupling, which is given by ($q = p'-p$)

$$\Gamma_\mu (p',p) = \bar{u}(p') [F_1(q^2)\gamma_\mu + \frac{i}{2m} \sigma_{\mu\nu} q^\nu F_2(q^2)] u(p).$$

For a point Dirac particle $F_1(q^2) = 1$ and $F_2(q^2) = 0$. The anomalous magnetic moment is given by $F_2(0)$. One of the earliest triumphs of QED was Schwinger's calculation of the magnetic anomaly to lowest order in α:

$$a_e = a_\mu = \frac{\alpha}{2\pi}$$

for which he received the Nobel prize.

Until recently, using the accepted value of the fine structure constant[1], one had

$$\frac{\alpha}{2\pi} = (1\ 161\ 409.835 \pm 0.244) \times 10^{-9}$$

A recent redetermination of the proton gyromagnetic ratio, which is needed for the conversion of e/h from the AC Josephson effect to α^{-1}, gives[2]

$$\frac{\alpha}{2\pi} = (1\ 161\ 410.039 \pm 0.128) \times 10^{-9}.$$

This obviously has consequences for the use of the lepton anomalous magnetic moments as tests of QED. The uncertainty is to be compared with the experimental[3] uncertainty in a_e:

$$a_e(\text{exp}) = (1\ 159\ 652.200 \pm 0.040) \times 10^{-9}.$$

The theoretical value of a_e is dominated by QED:

$$a_e^{QED} = \frac{\alpha}{2\pi} - 0.32847897\ (\frac{\alpha}{\pi})^2 + 1.1835(61)\ (\frac{\alpha}{\pi})^3 + 0(\frac{\alpha}{\pi})^4$$

$$= (1159\ 652.566 \pm 0.149) \times 10^{-9}$$

where the error in the α^3 term is an estimate of the uncertainty in numerical integrations for the 21 diagrams which have not yet been eva-luated analytically.

One should remark that the calculation of this contribution would not have been possible without the use of computers. Muonic and hadronic vacuum polarization contribute[4] (in units 10^{-9})

> Muon loop: 0.0028
> hadrons : 0.0016 ± 0.0002,

so that at present there is a slight discrepancy between theory and experiment

$$a_e(\text{theory}) - a_e(\text{experiment}) = (0.370 \pm 0.154) \times 10^{-9}.$$

Possible improvements involve the following:

a)   A better determination of the fine structure constant (not really theory) should be undertaken.

b)   It should be possible to reduce the uncertainty in the coefficient of $(\frac{\alpha}{\pi})^3$ by a factor of twenty. This needs new and better methods for doing numerical integrations[5]. Examples of graphs presenting special problems are shown in Fig. 1.

c)   A calculation of the contribution of order $(\frac{\alpha}{\pi})^4$ is in progress[4]. The 891 graphs involved group themselves into five classes

   i)   (25) second order vertex graphs with vacuum polarization (VP) loops. Examples are shown in Fig. 2a.

   ii)   (54) fourth order vertex graphs with VP loops, as indicated in Fig. 2b.

   iii)   (150) sixth order vertex graphs with a VP loop (Fig. 2c)

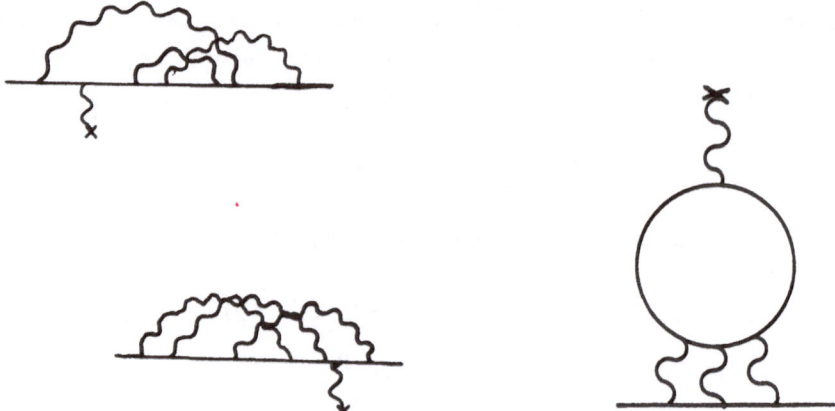

Fig. 1: Sixth-order contributions to g-2 which have not been evaluated analytically

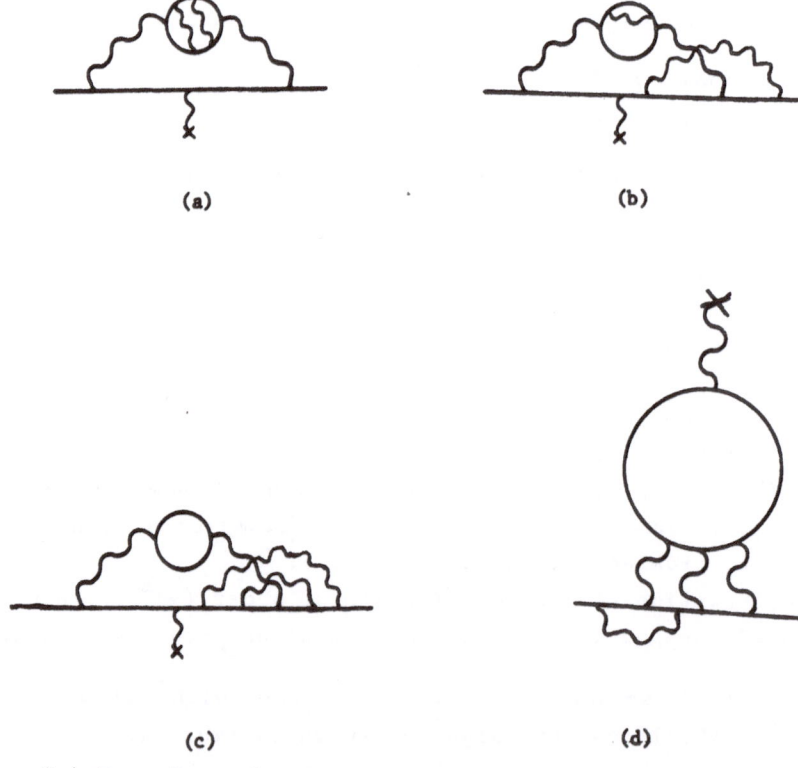

(a)   (b)

(c)   (d)

Fig. 2: Examples of graphs containing VP loops which contribute to g-2 in eighth order. See text for discussion.

iv) (144) light-by-light graphs with radiative corrections (Fig. 2d)

v) (518) graphs with no VP loops.

Kinoshita's calculation groups together the integrals for several related diagrams using the Ward identity. In this way the total number of integrals, having up to 10 dimensions, is slightly over 100. The integrands are generated by the program SCHOONSCHIP[6] and contain typically several thousand terms. The integrations are done on the CDC-7600 computer at Brookhaven National Laboratory. Kinoshita[7] reports that the diagrams of classes (i) + (ii) give a contribution

$$- 0.600 \ (\tfrac{\alpha}{\pi})^4.$$

He hopes that if all goes well, the rest will be computed to an accuracy of $\pm$ 10 % by August. The accuracy is limited by the computer budget.

It will not be necessary to say much about the muon anomaly since the subject will be reviewed by Dr. Farley[8]. Quantum electrodynamics predicts[9]

$$a_\mu(QED) = \tfrac{\alpha}{2\pi} + 0.765782 \ (\tfrac{\alpha}{\pi})^2$$

$$+ (24.45 \pm 0.06) \ (\tfrac{\alpha}{\pi})^3 + 135(64) \ (\tfrac{\alpha}{\pi})^4$$

$$+ \tfrac{1}{45} \ (\tfrac{\alpha}{\pi})^2 \ (\tfrac{m_\mu}{m_\tau})^2$$

$$= [1 \ 165 \ 852.4 \pm 2.4] \times 10^{-9} \ .$$

A contribution due to the weak interaction is estimated to be[9]

$$a_\mu(W.I.) \simeq 2 \times 10^{-9}$$

using the Weinberg-Salam model. Requiring that the weak contribution be less than the experimental uncertainty of $\pm$ 10^{-8} imposes some restrictions on models for the weak interaction.

The muon g-factor is also affected by hadronic vacuum polarization (Fig. 3,4). The main contribution (corresponding to Fig. 3) is given by

$$a_\mu(hadron) = \frac{1}{4\pi^3} \int_{4m_\pi^2}^{\infty} dt \ K_\mu(t) \ \sigma_{e^+e^- \to hadrons}(t)$$

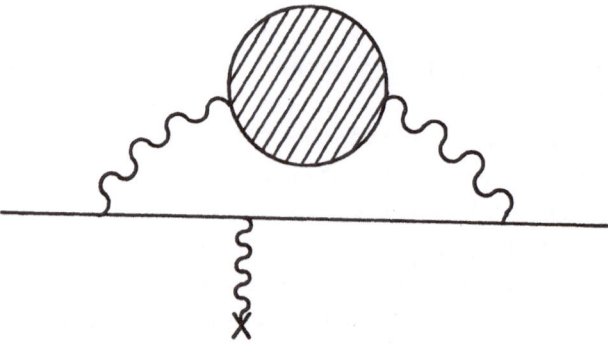

Fig. 3:   Lowest order hadronic VP

Fig. 4:   A higher order hadronic VP contribution which
          needs further study.

where

$$K_1(t) = \int_0^1 \frac{x^2(1-x)\ dx}{x^2+(1-x)t/m_1^2}$$

Inclusion of higher orders gives the result[9)]

$$a_\mu(\text{hadron}) = (66.7 \pm 9.4) \times 10^{-9}.$$

It should be possible to reduce the theory uncertainty by a factor of about 5. Already more data on the process $e^+e^- \to$ hadrons near the ρ-resonance and above 2.5 GeV exist so that a reevaluation of the lowest order contribution could considerably reduce the uncertainties associated with the lower order term (Fig. 3). The value of the contribution is unlikely to change much, however.

It will be more difficult to reduce the uncertainty in the contribution of higher order hadronic VP contributions (see Ref. 9). For one thing their evaluation is model dependent and the uncertainty due to this source can probably not be reduced much below the level

$$\Delta a_\mu(\text{hadron}) \approx \pm 10^{-9}.$$

This problem is particularly serious for the graph shown in Fig. 4. In fact, Kinoshita has even questioned the sign of this contribution. Further investigation is needed.

The total theoretical value (with the new value of α)

$$a_\mu^{(\text{th})} = (1\ 165\ 921.3 \pm 9.8) \times 10^{-9}$$

agrees nicely with experiment.

As previously mentioned the calculation of the magnetic anomaly of leptons has been made possible by the availability of powerful computers. Two different kinds of computing techniques are used in this field.

The first one allows the transformation of a given graph into a finite integral over several Feynman parameters. These methods belong to the area of symbolic and algebraic manipulations. Among the systems of programs which are used, because they can handle Dirac matrices, REDUCE and SCHOONSCHIP are probably the better known presently (see Ref. 10 for detailed references). This algebraic part of the computation of a graph usually takes only a few minutes of computing times, even at high orders of the perturbative expansion.

The techniques of the second kind which are required to evaluate the multidimensional integrals over the Feynman parameters are much more expensive. Although belonging to the area of numerical analysis, the best methods available have been set up by physicists[10]. The numerical integration of an n-tuple integral (n up to 10 at eighth order) may take several hours of CPU time on a powerful computer to get results with a good accuracy. Although the programs are based upon methods which have often yet to be proved mathematically correct, they have up to now given very precise results. Computer algebra is presently unable to calculate analytically such integrals. It seems that this situation will hold in the near future. The availability of symbolic integrators would be a step toward an improvement of the theoretical determination of the lepton anomalies.

References:

1.   Particle Data Group, Phys. Lett. 75B, 1 (1978)

2.   E.R. Williams and P.T. Olsen, Phys. Rev. Lett. 42, 1575 (1979)

3.   H.G. Dehmelt, this conference;
     R.S. Van Dyck, P.B. Schwinberg and H.G. Dehmelt, Bull. A.P.S. 24, 758 (1979).

4.   T. Kinoshita, CLNS-79/437. To appear in "Regards sur la Physique Théorique d'Aujourd'hui", H. Bacry ed., Editions du CNRS, Paris.

5.   M.J. Levine, E. Remiddi and R. Roskies, Phys. Rev. D20, 2068 (1979).

6.   H. Strubbe, Comp. Phys. Comm. 18, 1 (1979).

7.   T. Kinoshita and W.B. Lindquist, Cornell preprints CLNS-424 and CLNS-426 (1979).

8.   F. Farley, this conference;
     CERN-Mainz-Daresbury Collaboration, Nucl. Phys. B150, 1 (1979).

9.   J. Calmet, S. Narison, M. Perrottet and E. de Rafael, Rev. Mod. Phys. 49, 21 (1977).

10. J. Calmet and A. Visconti, to appear in "Field Theory, Quantization and Statistical Physics", E. Tirapegui edit., D. Reidel Pub. Co.

REFINED DATA ON ELECTRON STRUCTURE FROM INVARIANT FREQUENCY RATIOS IN ELECTRON AND POSITRON GEONIUM SPECTRA

Hans Dehmelt

Department of Physics, FM-15

University of Washington

Seattle, Washington 98195

USA

A new technique for observing the spectra of an individual electron trapped in a magnetic field and a parabolic axial electric potential (the geonium "atom") was demonstrated[1] by Van Dyck, Ekstrom and Dehmelt in 1976. This technique is based on the axial Stern-Gerlach Effect due to an auxiliary shallow magnetic bottle. The parabolic bottle field causes a slight dependence $\delta\nu_z = [m + n + 1/2 + (\nu_m/\nu_c)q] \cdot 1\,Hz$ of the axial oscillation frequency in the electric well, ν_z, on spin, cyclotron, and magnetron quantum numbers m, n, q. For example, a spin-flip from $m = -1/2$ to $m = +1/2$ changes the axial oscillation frequency $\nu_z \approx 60$ MHz by $\delta\nu_z = 1$ Hz. By monitoring ν_z spin, cyclotron and magnetron resonances may be detected by virtue of the changes in ν_z caused by them. In our experiment the axial frequency ν_z is known at all times as it is locked to a very stable rf source by means of a feedback circuit. This circuit also provides a dc signal proportional to shifts in the natural ν_z frequency which are nulled out by feeding the dc signal to the trap electrodes. The dc signal also is the principal observable in subsequent measurements. We induce spin flips not by applying a magnetic rf field at ν_s , the spin precession frequency, but by shaking the electron axially at $\nu_s - \nu_c$ through the inhomogeneous field of the magnetic bottle. This, in combination with the cyclotron motion at ν_c causes the electron to see an effective rf field at $\nu_c \pm (\nu_s - \nu_c)$ with the $\nu_c + (\nu_s - \nu_s)$ component causing the spin flips. The spin-cyclotron beat freqency $\nu_s - \nu_c$ is measured in this way. Cyclotron and magnetron resonances are detected by virtue of the large ν_z shifts occuring on resonance. From the measured ν_c and $\nu_s - \nu_c$ values ν_s and finally $\nu_s/\nu_c \equiv g/2$ are obtained. Actually, the observed cyclotron frequency value, now denoted ν_c', must be corrected for the electric field shift $\delta_e \equiv \nu_z^2/2\nu_c'$ to yield $\nu_c = \nu_c' + \delta_e$. The axial symmetry of the trap is checked by comparing δ_e with ν_m, the magnetron frequency. In an ideal trap $\delta_e = \nu_m$ should hold. In about 50 runs our measured frequency ratio

$$\nu_s/\nu_c = 1.001\ 159\ 652\ 200\ (40)$$

has been shown to be invariant under field variation from 18 to 51 kG. Uncertainties in δ_e contributed but a tiny fraction to our quoted error of 40 parts in 10^{12} which was predominantly due to an obsolete drifting magnet. Our ν_s/ν_c ratio which defines $g/2$ also equals the electron magnetic moment μ_s in Bohr magnetrons and is currently the most accurately known parameter of an elementary particle.

In analogous experiments on an individual positron,[2] Schwinberg, Van Dyck, and Dehmelt have measured the cyclotron frequency and compared it to that of an electron in the same field. This preliminary measurement of the ratio

$$\nu_c(e^-)/\nu_c(e^+) = 1.000\ 000\ 00\ (13)$$

which should equal the mass ratio $m(e^+)/m(e^-)$ constitutes a significant test of the CPT theorem for a charged elementary particle and its antiparticle. Experiments on the positron magnetic moment with expected error limits no larger than those realized for the electron are under way. A more detailed survey[3] of our work is available.

1.   R. Van Dyck, Jr., P. Schwinberg and H. Dehmelt, Phys. Rev. Letters 38, 310 (1977)

2.   P. Schwinberg, R. Van Dyck, Jr., and H. Dehmelt, Physics Letters, in press.

3.   H. Dehmelt, in ATOMIC PHYSICS 7, D. Kleppner and F. Pipkin, Editors, Plenum, 1981.

THE MUON MAGNETIC MOMENT

E. Klempt

Institut für Physik der

Universität Mainz, W. Germany

1. Introduction

Muon   mass   and   muon   magnetic   moment   are   not   calculable
within the theory of Quantum Electrodynamics; they rather define the
reference scale in which masses and energies should be determined when
muons are used to test QED or to search for anomalous muon couplings
which   could provide   the   "raison   d'être"   for   the   muon.   Only   the
anomalous part of the muon magnetic moment experimentally determined
with remarkable precision[1], can be calculated[2]. Theory and experiment
agree precisely once corrections being applied for hadronic and weak
processes which pollute the immaculated field of QED at high momentum
transfers. Muon   mass   and   muon   magnetic   moment   are,   in   particular,
needed for the interpretation of experiments which determine

- the muon spin precession frequency relative to its momentum
- the hyperfine splitting of muonium
- γ-transition energies in muonic atoms

In the following we will discuss the suitability of different experimental   methods   to   determine   the   muon   mass.   Measurements of the
total muon magnetic moment in units of the proton magnetic moment,
μ_μ/μ_p , will be considered as determinations of the muon mass because
the two quantities are related by

$$\frac{m_\mu}{m_e} = \frac{g_\mu}{2} \cdot \frac{\mu_B}{\mu_p} \cdot \left(\frac{\mu_\mu}{\mu_p}\right)^{-1}$$

where   the   constant $g_\mu \cdot \dfrac{\mu_B}{\mu_p}$ is   known[1,3]   with   a   precision   of
about 10^{-8}.   We   will   assume   that   CPT   holds   exactly,   so
that $m_{\mu+} = m_{\mu-}$ and $\mu_{\mu+} = \mu_{\mu-}$ .Because of their short life time muons
cannot   be   stopped   and   trapped   in   vacuum   as   it   has   been   done   with
electrons   and   positrons   extremely   successfully[4].   In - flight

experiments necessarily require large relativistic corrections; therefore muons have to be stopped in matter, when mass or total magnetic moment of muons is to be determined. Only the anomalous part of the magnetic moment is velocity independent and allows precision experiments using muons in flight.

Negative muons stopped in matter are captured by the Coulomb field of nuclei, they form muonic atoms. The γ radiation emitted in the de-excitation process can be used to determine the muon mass, but QED corrections (or other effects like nuclear finite size or electron screening) have to be taken into account in order to relate muon mass and the observed transition energies. Therefore, it is more appropiate to test QED corrections by measuring the γ radiation from muonic atoms[5], and to determine the muon mass in other experiments.

The hyperfine splitting in muonic atoms due to the interaction between muon magnetic moment and its orbital magnetic moment[6] or the magnetic moment of the electron cloud[7] depends on the muon magnetic moment, too. But also in these experiments μ_μ/μ_P should be regarded as input quantity.

Experiments determining μ_μ are generally easier and more precise than those determining m_μ , because muons are born polarized and the asymmetry of the decay positron distribution identifies the direction of the muon spin at the time of it's decay. Therefore, we restrict ourselves to a discussion of possible experiments aiming at a determination of the magnetic moment of positive muons μ_μ/μ_P . We will classify experiments according to following criteria:

- the magnetic field interacting with the muon magnetic moment can be applied externally, it can be the hyperfine field of a muon bound in muonium, or the superposition of hyperfine and external field,
- muons can reside in interstitial sites of a solid state lattice, they can be bound in muonium atoms, or in diamagnetic molecular bonds,
- the muon spin precession can be observed as free precession, by inducing radio frequency transitions between two energy levels with different expectation values of the muon magnetic moment, or by the Ramsey - Telegdi method.

2. Experimental Methods

2.1. The magnetic field

The muon magnetic moment can be determined by measurements of its interaction with magnetic fields. The magnetic field can be chosen as

- external magnetic field.

Then, muons acquire an additional energy of $H_i = -\vec{\mu}_\mu \vec{B}$ which corresponds to an energy $E = \pm \frac{1}{2} \mu_\mu B$ in energy eigenstates. A measurement of the energy difference $\mu_\mu B$ determines, of course, μ_μ / μ_p if the magnetic field is known from NMR measurements. The principal advantage of this method is the fact that the μ_μ defining relation is used to determine μ_μ .

- The hyperfine field of muonium at the position of the muon

leads to an energy splitting between the two states in which the total spin is one or zero, resp., which is proportional to the muon magnetic moment. But yet uncalculated higher-order QED corrections to the hyperfine splitting contribute an uncertainty of $\sim$ 1 ppm, and direct determinations of μ_μ / μ_p are therefore preferable. However, when

- hyperfine field and external magnetic field

are superimposed, a direct measurement of μ_μ / μ_p and a simultaneous determination of the hyperfine splitting become possible.

Fig. 1 shows the muonium hyperfine energy levels as a function of the external magnetic field. When two transitions, e.g. v_{12} and v_{34}, are measured in the same magnetic field, both, hyperfine splitting v_{HFS} and magnetic moment μ_μ / μ_p can be determined.

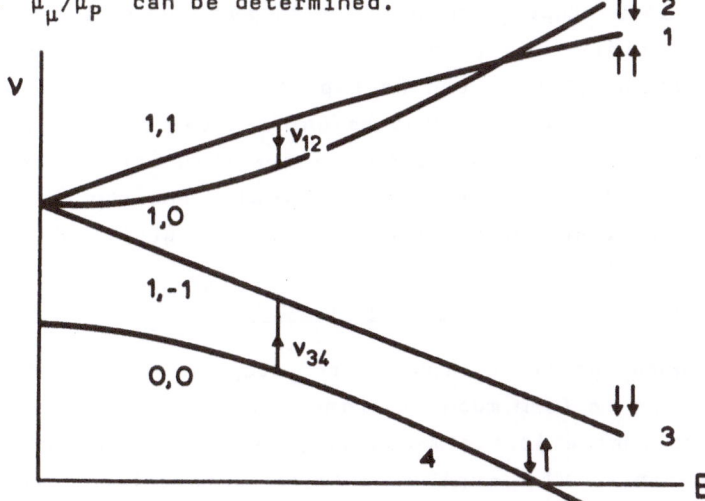

Fig. 1:
Hyperfine energy
levels of muonium

From Fig. 1 we see that

$$\frac{\delta v_{12}}{\delta B} = \frac{\delta v_{34}}{\delta B} = 0 \qquad \text{for} \qquad B = 11.3 \text{ kG}$$

Hence, the error in the determination of the average magnetic field in which muons precess propagates only weakly in to the final error of μ_μ/μ_p[8].

The use of the hyperfine field and an external magnetic field has the advantage that the transition or precession frequencies are well separated from those of free muons. Muons not stopping in the target vessel or its surroundings therefore do not contribute to the signal. Finally, we notice that for $B \sim 160$ kG $v_{12} = 0$.

2.2. The atomic status of muons

The atomic status of muons depends on the target in which they are stopped; it has decisive impact on the experimental techniques which can be used to determine μ_μ/μ_p .

- "Free" muon spin precession: solid state targets

The diffusion of hydrogen in metals is of great interest in solid state physics, and detailed investigations have been carried out[9]. It is found that hydrogen atoms disintegrate and that the proton diffuse with high rates within the metal lattice. Hence local magnetic field inhomogenities average out, and high-resolution NMR techniques can be applied. In nonmagnetic metals an external magnetic field will be slightly reduced or enhanced, an effect which is known as Knight shift.

Muons stopped in metals will average over field inhomogenities in a very similar way, and their Knight shift can be determined, too[10]. As isotope effects can be expected to be very small, a comparison of muon and proton spin precession frequencies in different metals should provide μ_μ/μ_p . But because of the skin effect, measurements of proton Knight shifts in metals are difficult, and the appropiateness of solids for precise measurements of μ_μ/μ_p has still to be investigated.

- Muons in muonium: gas targets

When positive muons are stopped in matter they can capture electrons to form muonium atoms. The effect of the muon magnetic moment on its energy levels makes it possible to determine μ_μ/μ_p . In order to minimize the muonium lattice interaction, muonium is formed in noble gases at low density. The residual effect of target gas on hyperfine structure interval and g_j factor are controled by varying the target gas pressure and allowing a linear and quadratic term in the pressure dependence.

- Muons in diamagnetic molecules: liquid targets

Muonium reacts with matter in a very similar way as hydrogen atoms. It can form stable diamagnetic compounds in which the muon magnetic moment experiences the external magnetic field which is only slightly reduced by the diamagnetic shielding; hence, the muon spin precession can be observed. A suitable liquid has to meet the following requirements:

it has to undergo fast chemical reactions with hydrogen, so that the muon is bound in a diamagnetic compound in a time shorter than the hyperfine structure period,

there should be one and only one chemical compound which is finally formed,

NMR measurements on the analogous hydrogen compounds must be possible,

the isotope effect of the diamagnetic shielding (i.e. the difference of diamagnetic shielding of a muon or a proton in analogous chemical compounds) must be calculable.

2.3. Techniques for muon spin precession detection

The technique for detection of the muon spin precession depends strongly on the direction of the initial muon polarization relative to the magnetic field. For

- muon spin polarization perpendicular to the magnetic field

the muon ensemble is represented by a coherent superposition of two energy eigenstates, and quantum beat oscillations can be observed. This technique assures the optimum statistical accuracy for a given number of muons. But in general it is limited in the rate of incoming muons: each decay positron must be attributed to its parent muon, and no second muon is allowed to enter the target between muon stop and decay. This limits the muon beam intensity to a few times 10^4 μ^+/sec while muon fluxes of several 10^6 μ^+/sec are available in muon factories. This rate limitation can be avoided only at particular accelerators and by use of appropiate methods. No such rate limitations exist if the

- <u>muon spin polarization is parallel to the magnetic field</u>

and radiofrequency transitions between two quantum states are induced
in which the expectation values of $-\mu_\mu B$ differ. Because of the skin
effect muons stopping in the microwave cavity or outside of it do not
contribute to the signal, and possible sources of background are eliminated. Part of the statistical power is lost, however, due to the
unavoidable power broadening of the line, since the power has to be
high enough to lead to a depolarization within the muon life time.

The maximum possible change in polarization depends on the choice
of the magnetic field: for a beam polarization of 1 it is

for external field · · · · for hyperfine field · · · · for hyperfine and
external field

2 · 0.5 · · · · · · · · · · · · · · · · · · 1

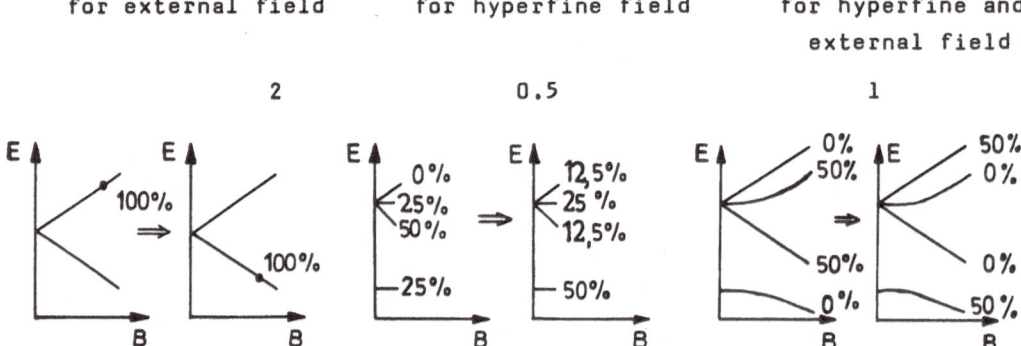

In the latter case, a double resonance technique[8] has to be employed
to reach the full statistical accuracy. These "figures of merit" for
the statistical power are, of course, effective also for other polarization detection techniques.

- <u>The Ramsey - Telegdi technique</u>[11]

of inducing a radiofrequency transition in two separated cavities or by
two separated rf pulses leads to narrower lines, in particular if the
separation in time of the two rf pulses is slightly larger than the
muon life time. An increase in statistical accuracy can thus
be achieved.

3. <u>Present Status</u>

There are three determinations of μ_μ/μ_P with experimental errors
of less than 3 ppm. In two experiments the precession of the muon
polarization in an external magnetic field was observed[12,13]. The
experiment of ref. 12 was performed at a conventional muon channel with

a limited flux of muons. Hence decay positrons can unambiguously be related to the last muon stop. Therefore the time interval between muon
stop and decay can be measured. Fig. 2 shows the number of detected
decay positrons as a function of this time interval. The oscillation
frequency can be determined from a fit to the data.

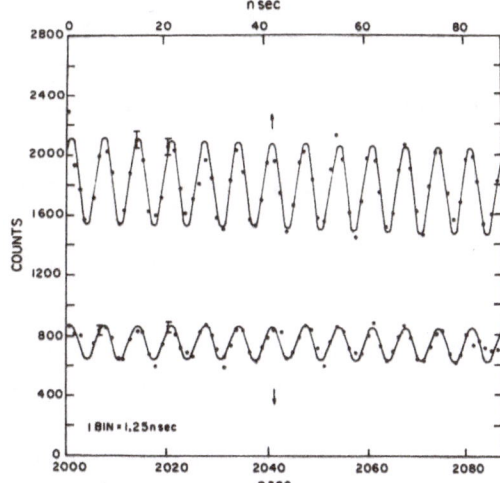

Fig. 2:
Differential time spectrum
of positrons from muon decay
(from ref. 12).

In the experiment of the ref. 15 the periodic time structure of
the SIN proton accelerator (50 MHz) was exploited by applying a
stroboscopic technique[14], where the muon precession frequency
coincided with the second harmonic of the beam burst repetition
frequency. There, maximum polarization builds up. Polarization is small
when the magnetic field is detuned (Fig. 3).

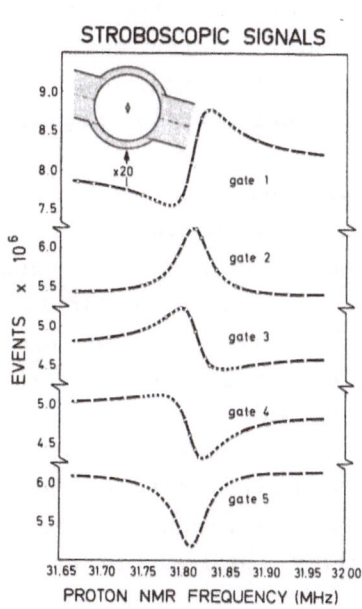

Fig. 3:
Stroboscopic signals as a function
of the magnetic field. Maximum polarization builds up when muon spin precession coincides with the frequency
of the stroboscope or its harmonics
(from ref. 13).

In both experiments muons were stopped in liquids: in H_2O and $CH_2(CN)_2$, or in pure Br_2 and in Br_2 contaminated with H_2O. As a result, the chemistry of muonium reactions in the epithermal and thermal domain can be considered as sufficiently understood, and the final chemical state of the muon is known.

In the experiment of ref. 13 muons neutralize to muonium atoms which form - within a time of less than 10^{-12} sec - muonium bromine, MuBr. If the Br_2 target is contaminated with H_2O the reaction chain

$$MuBr + H_2O \rightarrow (MuH_2O)^+ + Br^-$$

$$(MuH_2O)^+ + H_2O \rightarrow MuHO + (H_2O)^-$$

leads to formation of MuHO molecules with a reaction rate of about 10^{10} 1 M^{-1} sec^{-1}. In both molecules, MuBr and MuHO, the external magnetic field is reduced by the diamagnetic shielding. The shielding also depends on the neighbouring molecules, therefore it is different for MuHO in Br_2 and for MuHO in H_2O. This effect is known as liquid association shift.

Diamagnetic shielding and liquid association shift of protons in analoguous molecules (i.e. in HBr in Br_2; in H_2O in Br_2; and in H_2O) can be measured precisely in a conventional NMR spectrometer. The spin precession frequencies of muons stopped in Br_2, Br_2 contaminated with H_2O, and in H_2O follow the pattern of diamagnetic shieldings observed for protons in NMR measurements. This provides direct experimental evidence that muons indeed reside in those molecules which were deduced from chemical reaction arguments.

But muons are lighter than protons, and hence the zero point vibrations of muonium are larger than those of hydrogen. Therefore, the diamagnetic shielding of muons in molecules is smaller than that of protons, and also the liquid association shift may exhibit an isotope effect. In the case of H_2O and $CH_2(CN)_2$, only order of magnitude estimates for isotope effects exist, and the final error in μ_μ/μ_P was dominated by this uncertainty. On the other hand, quantum chemical ab - initio calculations for these isotope effects are available[15] for MuBr/HBr dissolved in Br_2 and MuHO/H_2O dissolved in Br_2. Their size is 0.5 - 1.0 ppm , and an error of less than 0.2 ppm was assigned to it. This error estimate is derived from the constraints on the quantum chemical calculations by experimentally known values, like equilibrium distance, vibrational energy and binding energy which have to be re-produced by the calculation.

The third measurement[16] was performed at Yale. Muons were stopped in a Krypton gas target at a pressure of a few atmospheres, and rf transitions between different eigenstates were induced. The presence of a strong magnetic field allowed the simultaneous determination of ν_{HFS} and μ_μ/μ_p . A magnetic field of 13.6 kG was chosen as at this field strength the two transitions ν_{12} and ν_{34} indicated in Fig. 1 have a ratio of 4/3 and can hence be driven by different excitation modes of one cavity. Fig. 4 shows a typical resonance line obtained by this method. Note that the line width is five times the natural line width because of power broadening.

Fig. 4:
Change of polarization
caused by resonant radio-
frequency transitions.

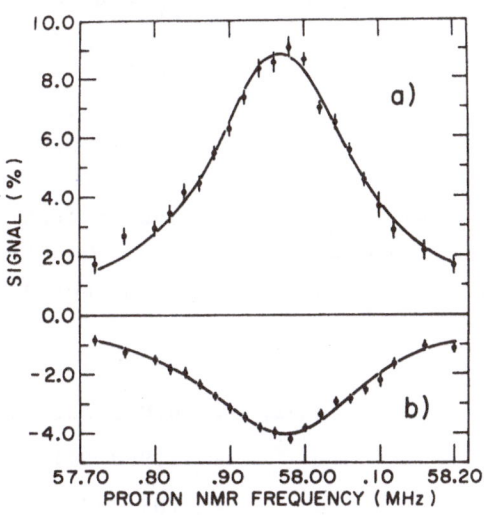

The three determinations of μ_μ/μ_p yield values of

$$\mu_\mu/\mu_p \; = \; 3.1833467 \quad (82) \quad \text{ref. 12}$$
$$= \; 3.1833441 \quad (17) \quad \text{ref. 13}$$
$$= \; 3.18334478 \quad (96) \quad \text{ref. 16}$$

Because of their excellent consistency, the average can be evaluated yielding

$$\mu_\mu/\mu_p \; = \; 3.18334464 \quad (83) \quad 0.26 \text{ ppm}$$

In the CERN Muon Storage Ring experiment[1] the anomalous part of the muon magnetic moment was determined to

$$\frac{\mu_a}{\mu_P} = 0.003707213 \quad (27)$$

With μ_μ/μ_P as given above the magnetic moment anomaly is calculated by

$$a_\mu(\text{exp}) = \frac{\mu_a/\mu_P}{\mu_\mu/\mu_P - \mu_a/\mu_P} = 0.001165923 \quad (8.5)$$

which is practically unchanged compared to the value given in ref. 1. It agrees excellently with the current theoretical result[2]

$$a_\mu(\text{theory}) = \frac{\alpha}{2\pi} + 0.765782 \frac{\alpha^2}{\pi^2} + (24.45 \pm 0.06) \frac{\alpha^3}{\pi^3} + 135 \frac{\alpha^4}{\pi^4} +$$

$$\frac{1}{45} \frac{\alpha^2}{\pi^2} \frac{m_\mu^2}{m_\tau^2} + (66.7 \pm 9.4) \cdot 10^{-9} + (2 \pm 2) \cdot 10^{-9} =$$

$$= 0.001165921 \quad (10)$$

The latter two contributions reflect the hadronic and weak contribution to the muon anomaly, resp. Here and from now onwards we use the following values for the fundamental constants[3]:

$$\alpha^{-1} = 137.035\ 963 \quad (15)$$

$$\mu_P/\mu_B = 0.001\ 521\ 032\ 209 \quad (16)$$

$$c = 2.997\ 924\ 58 \quad (1.2) \cdot 10^{19} \text{ cm/sec}$$

$$R_\infty = 109\ 737.314\ 76 \quad (32)\ \text{cm}^{-1}$$

Muon mass and muon magnetic moment are related by

$$m_\mu/m_e = \frac{g_\mu}{2} \cdot (\mu_\mu/\mu_P)^{-1} \cdot \mu_B/\mu_P =$$

$$= 206.76\ 8315 \quad (60)$$

The theoretical value for the hyperfine structure interval of muonium is

$$\nu_{HFS}(\text{theory}) = \frac{16}{3} \alpha^2 c\ R_\infty\ \mu_\mu/\mu_P\ (\mu_P/\mu_B)^{-1}\ (1+m_e/m_\mu)^{-3}$$

$$\{1 + \frac{3}{2} \alpha^2 + a_e + \epsilon_1 + \epsilon_2 + \epsilon_3 - \delta_\mu + \text{higher order terms}\}$$

$$a_e = \frac{\alpha}{2\pi} - 0.32847897 \frac{\alpha^2}{\pi^2} + 1.1835 \ (61) \ \frac{\alpha^3}{\pi^3}$$

$$\epsilon_1 = -\alpha^2 \left(\frac{5}{2} - \ln 2\right)$$

$$\epsilon_2 = -\frac{8\alpha^3}{3\pi} \ln\frac{1}{\alpha} \left(\ln\frac{1}{\alpha} + \ln 4 - 281/480\right)$$

$$\epsilon_3 = \frac{\alpha^3}{\pi} \ (18.4 \pm 5)$$

$$\delta_\mu = \frac{m_e}{m_\mu} \left[\frac{3\alpha}{\pi} \frac{m_\mu^2}{m_\mu^2 - m_e^2} \ \ln m_\mu/m_e - 2\alpha^2 \ \ln 1/\alpha \ (1 + m_e/m_\mu)^{-2} \right.$$
$$\left. + 2 \frac{\alpha^2}{\pi^2} \ (\ln m_\mu/m_e)^2 \right]$$

The resulting value

$$\nu_{HFS}(\text{theory}) = 4\ 463\ 303.62 \ (2.00) \ \text{KHz}$$

compares favourably with the most recent experimental value of

$$\nu_{HFS}(\text{exp}) = 4\ 463\ 302.91 \ (0.11) \ \text{kHz}$$

The theoretical expression for the muonium hyperfine structure interval contains the following errors:

From theory:	**1.24 kHz**
arizing from	
the error in ϵ_3 :	0.62 kHz
uncalculated parts in δ_μ :	
$\quad$ D $\quad m_e/m_\mu \ \alpha^2/\pi^2 \ \ln m_\mu/m_e \quad$ (D $\leqslant$ 3)	D $\cdot$ 0.14 kHz
$\quad$ E $\quad m_e/m_\mu \ \alpha^2 \quad$ (E $\leqslant$ 3)	E $\cdot$ 0.26 kHz
higher order corrections	$\sim$ 0.2 kHz
From experiments:	**1.56 kHz**
arizing from	
the error in μ_μ/μ_p	1.16 kHz
the error in α	0.98 kHz
the errors in other constants	0.34 kHz
Total error:	**2.0 kHz**

4. Is progress possible ?

Muonium is the only atomic system consisting of leptons only in which the effect of parity violating neutral currents might be accessible to experiments (experiments on positronium are less precise and the effect of P.V.N.C. on the hyperfine structure is smaller). P.V.N.C. will change the muonium hyperfine structure interval by about 10^{-8}; hence the theory of ν_{HFS} has to be completed to this level of precision. In addition, an equivalent precision would be required for μ_μ/μ_p .

I think it will be difficult to calculate the isotope effect of the diamagnetic shielding with a fractional accuracy of 10^{-2}, and that only low density noble gas targets make it possible to control lattice effects sufficiently precisely. But high flux muon surface beams of Arizona type can compensate the low stopping power of a gas target.

Technically one would like to combine

- a high magnetic field strength leading to a large number of precession cycles and
- a small precession frequency (100 MHz) so that the muon spin precession can be observed without excessive timing problems.

The two appearently contradictory requirements can be reconciled if an external magnetic field strength of 160 kG is chosen. At this field strength - which is just accessible by the present technology of super-conducting magnets or Bitter magnets - at this field strength the internal muonium hyperfine field and the muon spin remains fixed in space. Of course, it is impossible to inject low energy muons into a 160 kG magnetic field in a direction not parallel to the magnetic field lines. But muons can be stopped and a $\pi/2$ magnetic resonance pulse can be applied. Then, the muon spin precession in the plane perpendicular to the magnetic field can be observed. Rate limitations can be avoided at the muon channel of the booster synchroton of KEK, Tokyo, which is under construction[17].

In comparison to the previously discussed experiments this scheme provides two main advantages:

- the magnetic field is increased by a factor of 15
- the statistical power of the precession method is fully exploited.

Hence considerable improvements seem to be possible in the future.

References:

1. J.M. Bailey et al., Nucl. Phys. B150 (1979) 1
    See also F.J.M. Farley, "The Measurement of G-2 for the Muon",
    presented at this conference

2. See, e.g., E. Borie and J. Calmet, "The anomalous Magnetic Moment
    of the Leptons", presented at this conference, for a review.

3. Values of fundamental constants are taken from a recent compila-
    tion of B.N. Taylor, in: Review of Particle Properties, Rev. Mod.
    Phys. 52 (1980) S.1, Particle Date Group.

4. See, e.g., H.G. Dehmelt, "Invariant Frequency Ratios in Electron
    and Positron Geonium Spectra Yield Refined Data on Electron
    Structure", presented at this conference, for a review.

5. See, e.g., L. Tauscher, "Vacuum Polarization in Muonic Atoms",
    presented at this conference, for a review.

6. G. Carboni, G. Gorini, E. Iacopini, L. Palffy, F. Palmonari,
    G. Torelli, and E. Zavattini, Phys. Lett. 73B (1978) 229

7. P.A. Souder, D.E. Casperson, T.W. Crane, V.W. Hughes, D.C. Lu,
    H. Orth, and G. zu Putlitz, Phys. Rev. Lett. 34 (1975) 1417.
    See H. Orth, "Muonium and Related Topics", presented at this
    conference, for a review.

8. R. de Voe, P.M. McIntyre, A. Magnon, D.V. Stowell, R.A. Swanson,
    and V.L. Telegati, Phys. Rev. Lett. 25 (1970) 1779

9. G. Alefeld and J. Völkl (eds.): Hydrogen in Metals, Topics in
    Applied Physics, Vol. 28, 29 (Springer, Berlin, Heidelberg,
    New York 1978)

10. M. Camani, F.N. Gygax, W. Rüegg, A. Schenck, H. Schilling,
    E. Klempt, R. Schulze, and H. Wolf, Phys. Rev. Lett. 42 (1979) 679

11. D. Favart, P.M. McIntyre, D.Y. Stowell, V.L. Telegdi, R. de Voe,
    and R.A. Swanson, Phys. Rev. Lett. 27 (1971) 1336

12. K.M. Crowe, J.F. Hague, J.E. Rothberg, A. Schenck, D.L. Williams,
    R.W. Williams, and K.K. Young, Phys. Rev. D5 (1972) 2145

13. M. Camani, F.N. Gygax, E. Klempt, W. Rüegg, A. Schenck,
    H. Schilling, R. Schulze, and H. Wolf, Phys. Lett. 77B (1978) 326
    and "Measurements of the Magnetic Moment of the Positive Muon by a
    Stroboscopic Muon Spin Rotation Technique", submitted to Phys.Rev.

14. J. Christiansen, H.E. Manke, E. Recknagel, D. Riegel, G. Schatz, G. Weyer, and W. Witthuhn, Phys. Rev. Cl (1970) 613

15. M. Castro and J. Keller, "The Isotope Chemical Shift of μ^+ in HBr and H_2O", Report (1980) and to be published.

16. D.E. Casperson, T.W. Crane, A.B. Denison, P.O. Egan, V.W. Hughes, F.G. Mariam, H. Orth, H.W. Reist, P.A. Sonder, R.D. Stambaugh, P.A. Thompson, and G. zu Putlitz, Phys. Rev. Lett. 38 (1977) 956. The preliminary result of the final analysis of this experiment was presented in: H. Orth, "Muonium and Neutral Muonic Helium", this conference.

17. After this conference I was informed that the muon channel has started operation in July, 1980. T. Yamazaki, private communication

THE MEASUREMENT OF G-2 FOR THE MUON

F.J.M. Farley, The Royal Military College of Science,
Shrivenham, Swindon, UK.

The first part of the talk summarized the CERN experiments using a 3.1 GeV muon storage ring[1], also described in several review articles[2-4]. The essential elements of the experiment included

1. The equation for the spin precession relative to the momentum vector

$$f_a = (1/2\pi) \ a \ (eB/mc)$$
$$= af_s/(1+a)$$

where $a=(g-2)/2$, and the spin precession frequency at rest in the same field is

$$f_s = (1/2\pi)(eB/mc)(1+a)$$

2. The determination of f_s from the mean proton precession frequency f_p using the known value of the ratio $\lambda = f_s/f_p$

3. Lengthening of the muon lifetime by Einstein time dilation from 2.2 to 64 μs.

4. A continuous ring magnet, 14m in diameter with uniform field surveyed by proton nuclear magnetic resonance.

5. Electric quadruples inside the magnet to give vertical focusing.

6. Choice of the "magic" muon energy, at which the electric field does not affect the spin motion, given by $\gamma = (1+1/a)^{1/2}$.

7. Injection into the ring of a short (10ns) bunch of 3.1 GeV pions which decayed in flight leaving some forward polarized muons on stored orbits.

8. Decay of the muons in flight with the decay electrons emerging
on the inside of the ring giving a counting rate modulated by the
spin motion.

9. Observation of the rotation frequency of the muon bunches at
early storage times to establish the radial distribution of the
particles in the magnet aperture.

The result was $\quad R \equiv f_a/f_p = 3.707213(27)$ (7.2ppm).

Using the mean of the best three available measurements[5-7] of λ,

$$\lambda = 3.1833437(33)(0.7\text{ppm})$$

this gives $\qquad a = R/(\lambda-R) = 1165924(8.5) \times 10^{-9}$

in excellent agreement with the theoretical value

$$a^{th} = 1165921(8.3) \times 10^{-9}.$$

It is remarkable that the experiment is thus sensitive to the
interaction energy between the anomalous magnetic moment and the
magnetic field, $a(eB/mc)\hbar$ at a level of 7 ppm, i.e. to 10^{-11} eV, while
the muon energy in the laboratory is 3 GeV, a ratio of over 10^{20}.

The following conclusions may be drawn:

1. The QED calculations of the muon anomaly are confirmed to 4.7%
in the sixth-order term.

2. For conventional modifications of the photon propagator the
cut-off parameter Λ_γ must be a least 21 GeV.

3. The muon behaves as a point charge. If there is a form factor
$F(q) = 1-q^2/\Lambda_\mu^2$, then $\Lambda_\mu > 36$ GeV.

4. The hadronic vacuum polarization contributes $(67\pm10)\cdot10^{-9}$ compared with the theoretical $(66\pm10)\cdot10^{-9}$, and is thus established
at a level of 5 standard deviations.

5. No new couplings of the muon are apparent.

6. No undiscovered lepton of mass less than $2 \times$ (muon mass) can
exist, as this would give a detectable further contribution to the
vacuum polarization.

7. The electric dipole d_μ moment of the muon was checked in a special variation of the experiment with the result $d_\mu = (3.7\pm3.4)\cdot10^{-19}$ e·cm.

8. The dilated lifetime was measured in a special run with careful control of the muon losses and confirmed the theory of relativity to 0.1% at $\gamma = 30$. It is emphasized that this is in a circular orbit corresponding to the conditions of the controversial "twin paradox".

Possible new experiment

Is there any way of improving the accuracy of the experiment by a further factor of 15? This would enable the contribution of the weak interactions to be detected. The Weinberg-Salam theory predicts about 2ppm in a, due to virtual $W^\pm$ and Z^0, but the number is very sensitive to the cancellations in the model, and variants of the theory give widely different results. In spite of the success of the gauge theories there is at present no evidence for the existence of these particles. A measurement of the muon anomaly to 0.5 ppm would not only confirm their existence but give important guidance as to the correct version of weak interaction theory.

The programme would require

a.   a better calculation of the QED terms: present error 1.7 ppm,

b.   better data on $\sigma_{e^+e^- \rightarrow \text{hadrons}}$ leading to a better estimate of hadronic vacuum polarization,

c.   a better value of $\lambda = f_s/f_p$: present error 0.7 ppm.

All these steps would seem to be achievable.

To improve the accuracy of the (g-2) measurement a strong focussing ring could be used with increased Bγ giving more (g-2) cycles per muon lifetime, and a larger number of stored particles. The CERN ISR has been considered as an example.

A fundamental difficulty in a strong focussing machine concerns the calibration of the magnetic field. This can be solved in principle by sending transversely polarized deuterons round the same orbits. It is assumed that their polarization as a function of time can be followed using a nuclear interaction with a gas jet target. In effect the (g-2) precession of the deuteron is used to calibrate the magnet.

One has

$$f_a^\mu = \gamma^\mu \, a^\mu \, \beta^\mu \, f_o \tag{1}$$

$$f_a^d = \gamma^d \, a^d \, \beta^d \, f_o \tag{2}$$

where f_a^μ and f_a^d are the muon and deuteron (g-2) frequencies, γ^μ, γ^d are the relativistic energy factors, a^μ, a^d the anomalous magnetic moments, β^μ, β^d their velocities and $f_o = c/2\pi R$ defines the orbit radius R assumed to be identical. As the momenta are the same

$$\beta^d \gamma^d m^d = \beta^\mu \gamma^\mu m^\mu, \tag{3}$$

so

$$a^\mu = a^d \, (f_a^\mu/f_a^d) \, (m^\mu/m^d) \tag{4}$$

In this equation m^μ is known to 0.7 ppm from the value of λ, m^d to 0.05 ppm and a^d to 2.3 ppm, but this latter could readily be improved. f_a^μ and f_a^d would be measured.

The orbits of muon and deuteron could be defined by timing the rotation frequencies. But the momentum compaction factor α means that

$$\Delta R/R = \alpha(\Delta p/p) = \alpha(\Delta\gamma/\gamma) \tag{5}$$

so the orbit frequency (radius) must be measured more accurately than the desired accuracy in γ.

In the ISR $\alpha = 0.011$, so to fix γ to 0.2 ppm would mean defining time to 10^{-3} ns over a base of 500 µs; may be just possible <u>in average</u> over many particles. A machine with $\alpha \sim 0.1$ would be better.

If a short bunch of particles is injected and followed for say N turns it spreads out in time because of the spread in momentum (radius). This takes longer than in a weak focussing machine, again because of momentum compaction, so it will be a long time before the bunches overlap in time. In the ISR at 30 GeV, bunches initially 105 ns apart would overlap after 500 µs for $\Delta p/p = 1.8\%$, compared with a muon life time of 640 µs.

Inside a particular bunch the time of arrival of a particle correlates with γ; the spin angle also correlates with γ so one would see the (g-2) precession <u>inside the bunch</u> at an apparent frequency which can be deduced from eqns (1) and (5)

$$f_a' = \alpha \, \gamma(\beta c/2\pi R)/(\alpha-\gamma^{-2}) \tag{6}$$

This result applies to both μ and d using the corresponding parameters. Note that the effect of the momentum compaction is to make the (g-2) modulation inside each bunch $(\alpha - 1/\gamma^2)^{-1}$ times faster than normal.

For the ISR we find that the real (g-2) period followed by individual muons and deuterons T, and the apparent period T' seen inside the bunches, are as follows:

	a	γ	T μs	T' ns
μ	.00116	286	9.46	104
d	-0.143	16.1	1.36	9.7

The stored muon intensity has been estimated for the ISR, assuming injection of 30 GeV pions from the SPS. With the horizontal aperture stopped down to 3 cm, Δp/p is 1.8%, and the stored muon intensity about 73k per SPS cycle, with typically a decay electron counting rate of 5k per cycle, 25 times greater than the previous experiment. With the average magnetic field B̄ down to 0.5 but γ up by 10, this suggests statistical accuracy 25 times greater for the same number of machine cycles.

The possibility of using the ISR in this way seems remote. Perhaps it would be better to design a new high-field strong focussing ring with parameters such as:

    field   6 T
    stored energy   6 GeV
    orbit diameter   5 m
    α   0.1
    muon lifetime   128 μs
    decay electron rate (say) 400 per PS cycle.

With Bγ up by a factor 10 the statistical accuracy would be more than sufficient, and the timing to about 0.01 ns on average is feasible. If calibration with polarized deuterons can be achieved this would make an attractive experiment for the CERN proton synchrotron.

References:

1) Bailey, K. Borer, F. Combley, H. Drumm, C. Eck, F.J.M. Farley, J.H. Field, W. Flegel, P.M. Hattersley, F. Krienen, F. Lange, G. Lebee, E. McMillan, G. Petrucci, E. Picasso, O. Runolfsson, W. von Rüden, R.W. Williams and S. Wojcicki, Nuclear Physics B150, 1 (1979) 1-75.

2) F.J.M. Farley and E. Picasso. Ann. Rev. Nucl. Part. Sci. 29, 243 (1979)

3) F.H. Combley, Reports on Progress in Physics, 42, 1889 (1979)

4) F.H. Combley, F.J.M. Farley and E. Picasso, Physics Reports 68, 94 (1981)

5) K.M. Crowe, J.F. Hague, J.E. Rothberg, A. Schenck, D.L. Williams, R.W. Williams and K.K. Young, Physical Review D5, 2145 (1972)

6) D.E. Casperson, T.W. Crane, A.B. Denison, P.O. Egan, V.W. Hughes, F.G. Mariam, G. Orth, H.W. Reist, P.A. Souder, R.D. Stambough, P.A. Thompson and G. zu Putlitz, Physical Review Letters 38, 956 (1977)

7) M. Camani, F.B. Gygax, E. Klempt, W. Ruegg, A. Schenck, H. Schilling, R. Schulze and H. Wolf, Physics Letters 77B, 326 (1978)

QUANTUM ELECTRODYNAMICS IN BOUND SYSTEMS

E. BORIE

Institut für Theoretische Kernphysik der
Universität Karlsruhe, W. Germany

Experimental aspects of QED in bound systems will be treated in detail in other talks at this conference. I shall therefore attempt more to give a general theoretical introduction to the subject, without going into a detailed comparison between theory and experiment, rather than to give a comprehensive review of the subject.[1] I shall thus attempt to explain which QED effects can be investigated best with bound systems, some of the theoretical problems which one encounters in calculating QED effects in bound systems (although the list of problems will surely be incomplete) and to give some (subjective) suggestions as to where further improvements are desirable and/or possible.

The Bethe-Salpeter equation is usually taken as the starting point for the relativistically covariant description of a bound system. In practice, it is necessary to approximate the Bethe-Salpeter equation by a potential equation which one can solve. The procedure for doing this is not unique; the separation of the problem into lowest order approximation and corrections can be done in several different ways. A number of alternatives exist in the literature [2-6] and there is no way to tell which of them is the "correct" (or at least the most useful) one. The question of what is the best relativistic two body equation for the calculation of bound states is obviously of great interest, not only for QED, but also for other bound systems like baryonium or charmonium. It is also interesting to inquire whether the different approaches which have been proposed are equivalent to all orders of perturbation theory, or whether at some point the entire procedure breaks down in the sense that different approaches give different results, even in the absence of calculational errors. Investigations in muonium and positronium can provide sensitive tests of these questions.

For the case of normal or muonic atoms, the most suitable approach is to treat the heavy nucleus as nonrelativistic and on the mass shell. One then obtains as a first approximation for the description of the lepton's motion the Dirac equation with reduced mass. One thus regards the lepton as a point particle having spin 1/2 which interacts with a given external field. There are two types of corrections to this picture. In atomic physics, one class is known as the relativistic recoil corrections. A part of these can be very simply obtained from a relativistic generalization of the reduced mass and a more careful definition of the Sommerfeld parameter.[6,7] The remainder are a consequence of the fact that the Dirac equation with reduced mass is only an approximation to the fully covariant description of a two body system. The leading corrections, known as the Breit and non-Breit corrections, are typically of order $(\alpha Z)^n$ m_-/m_+ ($n = 1,2,..$) relative to the uncorrected binding energies.

The relativistic recoil corrections would be present even in the absence of the quantized radiation field. The corrections which arise primarily from the interaction between the lepton and the quantized electromagnetic field can be described as "pure QED" effects. The radiative corrections are due to the emission and/or absorption of real or virtual photons. As a result of this, the lepton acquires form factors and no longer behaves precisely as a point Dirac particle. Vacuum polarization results in a modification of the photon propagator due to virtual pair production and re-annihilation; this leads to a modification of Coulomb's law at short distances (less than the Compton wavelength of the particle produced). The experimental consequences of radiative corrections or vacuum polarization include, among others, the lepton anomalous magnetic moment, shifts in atomic binding energies and annoying backgrounds for electron scattering experiments.

Before going into a more detailed discussion of experimental tests of QED and recoil effects in bound systems, it may be useful to describe the pure QED effects in a little more detail and indicate some of the non-QED effects which one must take into account in order to make a precise comparison between theory and experiment.

1.   Self energy and vertex graphs

In normal atoms the most important QED effect is the self energy or, more precisely the fact that the self energy of a bound electron is not the same as the self energy of a free one. The difference is finite and observable (Lamb shift).

The calculation of the self energy correction provides an excellent example of how simple perturbation expansions may not always be appropriate. The expansion parameter αZ occurs in the lepton wave functions, and thus also in the lepton propagators; it also appears in the Coulomb and transverse photon propagators. Fig. 1 shows the reduction of the self energy graph in the bound interaction picture to a set of vertex graphs (here double lines represent lepton wave functions or propagators in the presence of the external field, single lines in the absence of the external field as, for example in ref. 8). If one continues the expansion in powers of $V \sim \alpha Z$, one discovers that the extra factors αZ can be compensated by factors p^{-1} where $p \sim m \alpha Z$ in the momentum integrals. Terms of the same order in αZ also come from relativistic effects in lower order parts of the expansion and from the momentum dependence of the lepton form factors. Also terms involving $\ln(\alpha Z)^{-1}$ appear, indicating that an expansion about the point $\alpha Z = 0$ does not have nice analytic properties.

Thus an expansion in terms of αZ is anything but straightforward. In fact it does not seem possible to specify a set of rules which is guaranteed to generate all terms of a given order and the methods which one uses are not unique when one tries to extend them to higher orders.

This problem is well illustrated in the case of the contribution to the Lamb shift in order $(\alpha Z)^6 m_e$. Two calculations [9,10] of these higher order binding corrections exist; the results differ by more than the experimental uncertainty and the question of which is correct is still open. Another case in which these difficulties have only recently been resolved is the case of corrections of order $(m_-/m_+) \alpha^4 \ln \alpha^{-1} R_\infty$ in the hyperfine structure of muonium and positronium.

2.   Vacuum polarization

The virtual creation and annihilation of e^+e^- (or other particle-antiparticle) pairs gives rise to a modification of Coulomb's law at distances small compared with the electron's Compton wavelength $\lambdabar_e \simeq 386$ fm. The virtual pair modifies the electric field produced by a given charge distribution; this is analogous to classical electrostatics in a medium. The observed field is the sum of the fields produced by the "true" and "polarization" charges. The separation is similar on a qualitative level for vacuum polarization, although the mathematical details differ. The result of a more detailed calculation

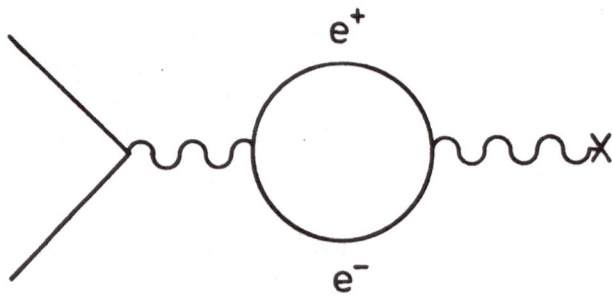

1. Decomposition of the self-energy diagram.

2. Graph giving rise to the Ühling potential.

is that at distances large compared with λ_e the "true" nuclear charge is screened by the virtual electron. This screened charge is by definition the observed nuclear charge Ze since this is the charge which produces the nuclear Coulomb field at macroscopic distances. As a test charge is brought nearer it effectively "sees" a larger charge and the electric field is increased in comparison with a pure Coulomb field. This will result in an increase in atomic binding energies. For calculational purposes, one usually describes the effect of vacuum polarization by the Ühling-Serber potential [11,12] (see fig. 2) and treats Coulomb corrections to the electron propagator (fig. 3)[13] and higher order terms (fig. 4)[14] as additional perturbations.

The effect of vacuum polarization is rather small compared with that of self energy for normal atoms, since the Bohr radii are of the order of 137 Z^{-1} λ_e. However, the effect of vacuum polarization is the dominant QED effect in muonic atoms, since the Bohr radii are of the order of 0.66 Z^{-1} λ_e. In fact for this case, the higher order corrections are also numerically quite important.[15]

In tests of QED with atoms, there are further effects which can disturb the measurement as a QED test. Among the most important of these are those due to nuclear size and structure.

For the description of atoms, the nucleus is approximated as the source of a static external potential arising from an extended charge distribution. The nuclear charge distribution is measured by electron scattering, although not always accurately enough. This effect by itself shifts the energy levels from their point Coulomb values. Naturally, s-states are most strongly affected.

In addition, the nuclear extension also has an effect on the wave function of the bound lepton and on the operators (effective potentials) whose expectation values give the radiative corrections. The latter effect is illustrated in fig. 5. This will mix the QED and nuclear effects.

In reality, the nucleus is not simply the source of an electrostatic potential; it has its own internal degrees of freedom and excitation spectrum. The correction due to the virtual excitation of these degrees of freedom is known as nuclear polarization. For most atomic orbits of interest for QED tests, this correction can probably be estimated accurately enough from second order perturbation theory

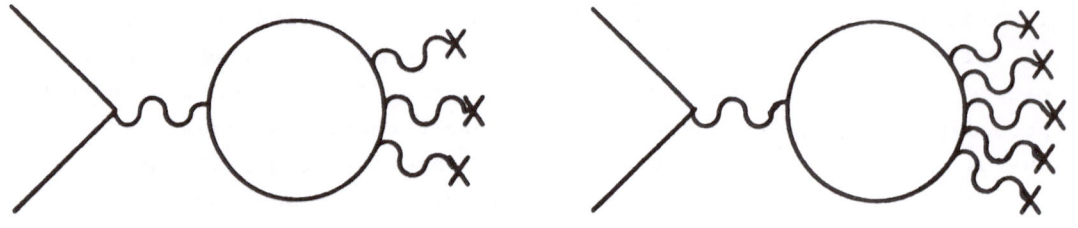

3. Coulomb corrections to the electron propagator in the vacuum polarization

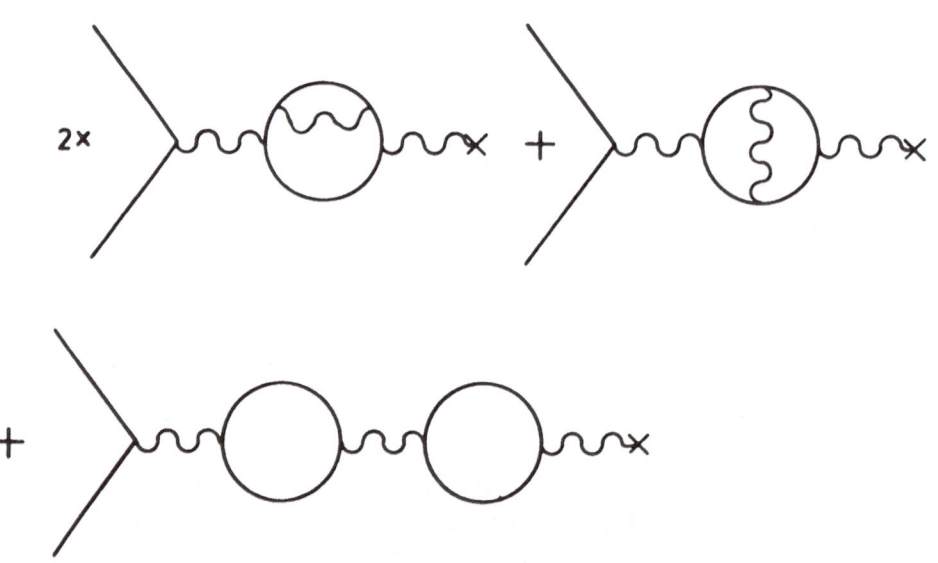

4. Fourth order vacuum polarization diagrams.

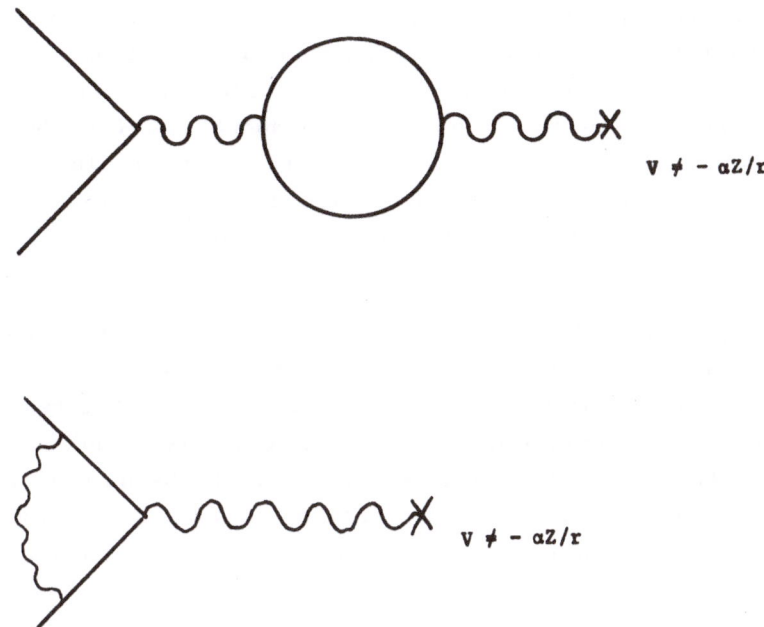

$v \neq - \alpha Z / r$

$v \neq - \alpha Z / r$

5. Mixing of radiative corrections and nuclear size effects.

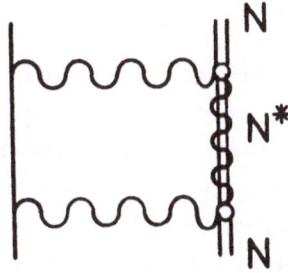

6. Nuclear polarizability. The intermediate nucleus is excited.

(fig. 6). The calculation involves a nuclear model which must be adjusted to fit measured photoabsorption cross sections. If this is done
carefully, it should be possible to calculate the nuclear polarizability correction to an accuracy of about 10% for the cases of interest here (this does not apply to low-lying levels of heavy muonic
atoms). It is probably not worthwhile to measure a transition energy
more accurately than about 10% of the nuclear polarization correction
in most cases.

The polarizability correction is the limiting factor in using the
hydrogen ground state hyperfine structure as a test of QED. The
measurement is accurate to 2×10^{-6} ppm while the estimated value of the
polarizability correction is about 1 ppm. It might be possible to
calculate the correction to an accuracy of 20% using multipole analysis
of pion photo- and electroproduction in the resonance region as well as
data on deep inelastic scattering of transversely polarized electrons
by protons. Experiment would still be five orders of magnitude ahead of
theory!

In the case of muonic atoms, interaction between the muon and the
remaining atomic electrons also has an influence on the energy levels
which cannot be neglected.

Other factors which are important when discussing tests of QED
with bound systems are the accuracy with which the parameters of the
theory are known. This was clearly illustrated in the case of the a-
nomalous magnetic moment of the electron; the main source of "theoretical" uncertainty is the numerical value of $\alpha/2\pi$.

I shall now try to give an overview of some of the more recent
theoretical work and examples of where improvements are most badly
needed. Unfortunately (for theorists) experiment is way ahead of theory
in most cases and improvements in the calculations will be difficult. I
shall take the standpoint that a meaningful test of QED must be
sensitive to radiative or relativistic recoil effects at a nontrivial
level. Thus for example the fine structure of the 2p level in hydrogen
does not really provide us with a test of QED since it is dominated by
the prediction of the Dirac theory with radiative effects (other than
the electron's anomalous magnetic moment) entering only at the level of
about 1 ppm.

Muonium and positronium are "pure" QED systems in that there is
almost no contamination from other interactions. They also provide the

most sensitive test of relativistic quantum two-body equations.

Observables are the ground state hyperfine structure, fine and hyperfine structure of excited states (mainly the n = 2 level) and positronium lifetime.

The ground state HFS is dominated by the Fermi formula:

$$\nu_F = \frac{\mu_+}{\mu_e} \frac{16}{3} Z^4 \alpha^2 R_\infty \left(\frac{m_r}{m_e}\right)^3 (1 + a_e) \;.$$

We then have

$$\nu = \nu_F (1 + \delta_{QED} + \delta_{rec}) \;.$$

δ_{QED} is known up to terms in α^3/π with an uncertainty of $\pm 5 \; \alpha^3/\pi \approx 0.62$ ppm.[17] For positronium, there is an additional contribution from virtual annihilation (fig. 7).

In the case of muonium, the experimental[18] hyperfine interval

$$\nu_{exp} = 4463302.35 \pm 0.52 \text{ kHz}$$

is known a factor ten more precisely than the theoretical predictions. The theory uncertainty is largely due to the uncertainty in the parameters α(1 kHz) and μ_μ/μ_p, although progress has been and is continuing to be made.

For muonium,

$$\delta_{QED} = - (2.340 \pm 0.062) \; 10^{-5}$$

$$\delta_{rec} = \frac{m_e}{m_\mu} \left[-3 \; \frac{\alpha}{\pi} \; \frac{\ln(m_\mu/m_e)}{1-m_e^2/m_\mu^2} + \frac{2\alpha^2 \; \ln\alpha^{-1}}{(1+m_e/m_\mu)^2} - 2\left(\frac{\alpha}{\pi}\right)^2 \ln^2(m_\mu/m_e) + \cdots \right]$$

Improvements in theory since 1977 include the correct coefficient of $\alpha^2 \ln \alpha^{-1}$ (refs. 2-4) and the terms[19] in $(\alpha/\pi)^2 \ln^2(m_\mu/m_e)$ in δ_{rec}. The latter contribute -6.6 kHz, while the uncalculated terms are expected to contribute 3-4 kHz. Other theory uncertainties include ± 2 kHz from the uncertainty in δ_{QED} and the uncertainty in the value of the muon's magnetic moment. Using the recent, as yet unpublished values quoted by Dr. Klempt at this conference, as well as the most recent value for the fine structure constant[20], one finds

$$\nu_{th} = 4463303.2 \pm 4.4 \text{ kHz if } \mu_\mu/\mu_p = 3.1833441 \text{ (17) (0.5 ppm)}$$

$$= 4463308.4 \pm 5.5 \text{ kHz} \qquad = 3.1833478 \text{ (26) (0.8 ppm)}$$

A previously mentioned discrepancy between theory and experiment has been eliminated, mainly by the improved calculation of δ_{rec}. Improvements in theory will require a more accurate evaluation of the coefficient of α^3/π in δ_{QED}, calculation of the terms in $(\alpha/\pi)^2 \ln(m_\mu/m_e)$ and α^2 in δ_{rec} and more accurate measurement of the parameters of the theory.

The hyperfine structure of neutral muonic helium has been measured[21]. The theory needs a better three-body treatment of the main term. The radiative corrections have been calculated to sufficient accuracy[22].

The calculation of the energy levels of positronium differs from that for muonium in several respects. Since the masses are equal, one should not use the Dirac equation even as a first approximation, but is rather forced to use a relativistic two body equation. Also the fine structure and hyperfine structure are comparable in magnitude. Furthermore, virtual annihilation diagrams (fig. 7) are quite important. The spectrum of positronium is shown in fig. 8. The spectroscopy of the n-2 levels is discussed by Weber.[23] Here I shall discuss only the ground state hyperfine splitting. There are two experimental values quoted in the literature[24,25].

$$\nu_{exp} = 203384.9 \pm 1.2 \text{ MHz}$$

$$= 203387.0 \pm 1.6 \text{ MHz},$$

to be compared with a theoretical value of 203400 ± 10 MHz. The slight discrepancy between theory and experiment should not be taken seriously since not all terms of order $\alpha^4 R_\infty$ have been calculated. Following Fulton[26] one has

$$\nu = \nu_Q + \nu_R + \nu_A$$

where

$$\nu_Q = \frac{2}{3} \alpha^2 R_\infty [(1 + a_e)^2 + \delta_{QED}]$$

$$\nu_R = \alpha^2 R_\infty [-\alpha/\pi + C_R \alpha^2 \ln \alpha^{-1} + B_R \alpha^2 + \ldots]$$

$$\nu_A = \alpha^2 R_\infty [1/2 + 2.1376 \alpha/\pi + C_A \alpha^2 \ln \alpha^{-1} + B_A \alpha^2 + \ldots]$$

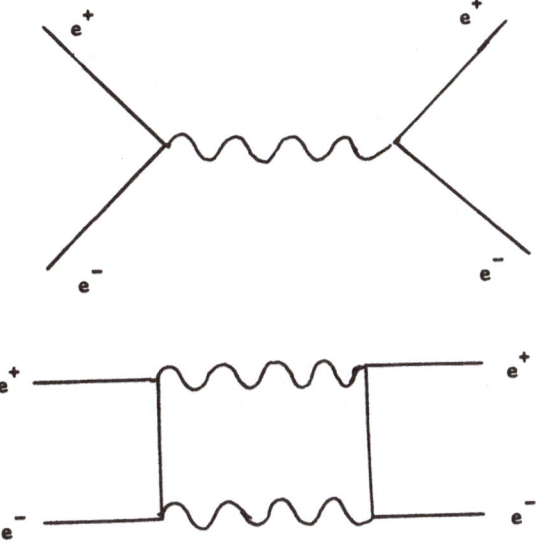

7. Examples of virtual annihilation graphs which contribute to the fine
    and hyperfine structure in positronium.

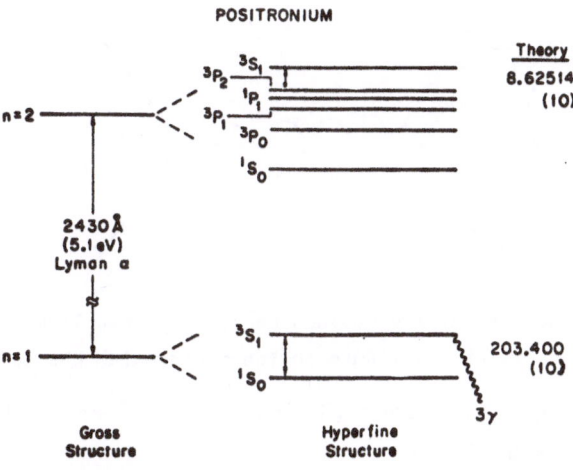

8. Positronium spectrum.

Here δ_{QED} is calculated as for the hyperfine structure of hydrogen and muonium[26]. In order to see that the coefficients B_R and B_A must be fully calculated before a meaningful comparison with theory is possible, I remark that the hyperfine interval ν has been measured with a precision of about 1.5 MHz while $\alpha^4 R_\infty$ = 9.3 MHz.

Only recently have all contributions of order $\alpha^4 R_\infty \ln \alpha^{-1}$ been determined. As the table indicates, the contribution has been strongly time dependent. The difficulty has been the proper treatment of Coulomb photons in the radiative corrections (see fig. 9). Since these photons are also responsible for binding, there is a danger of double counting, or of leaving something out if the contribution of the Coulomb photons is subtracted. By now, two groups[3,4] agree that all contributions of this order have been found and the result may be assumed to be correct.

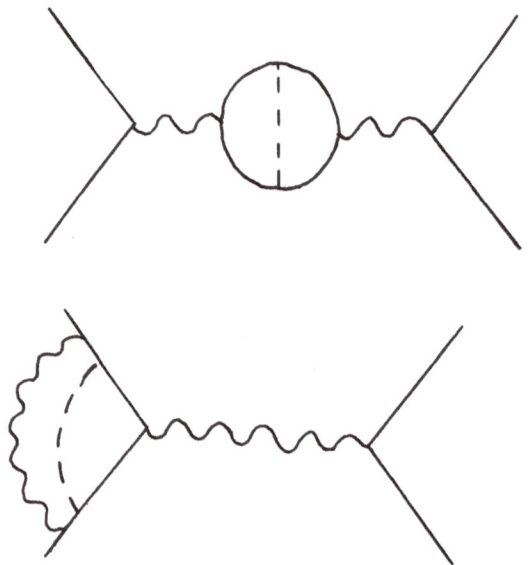

9. Some diagrams containing Coulomb photons (dashed lines) as binding corrections, which contribute to the positronium HFS in order $\alpha^4 R_\infty \ln \alpha^{-1}$.

Some of the contributions of order $\alpha^4 R_\infty$ have been calculated. These include, among others:

the contribution to ν_Q -18.1 MHz

contributions from two and three photon annihilation[27] +12.2 MHz.

Several independent groups are working on the remaining contributions to this order and estimate that final results should be forthcoming in about one year.

Table 1: Time dependence of the $\alpha^4 \ln \alpha^{-1} R_\infty$ contribution to the ground state hyperfine splitting in positronium.

Year	$C_R + C_A$	Contribution in MHz
1971	3/4	34
1973	1/2	23
1976	7/12	27
1978	5/12	19

It is well known that positronium in the singlet s-state decays to two photons with a lifetime which corresponds to a width of 5 μeV, making it the narrowest e^+e^- resonance known. The theoretical value of the lifetime, including contributions of relative order $\alpha^2 \ln \alpha^{-1}$ is[28]

$$r_{1_{S_0}} = 7.9867 \; (ns)^{-1} \; .$$

In this case, the experimental[29] lifetime of (7.99 ± 0.11) ns is not known precisely enough even to test the corrections of relative order α/π although possible improvements are being considered.[30]

Positronium in the triplet s-state decays to three photons and precision measurements have been made for this case. These are reviewed in ref. 30. Here I simply indicate in fig. 10 how both the theoretical predictions and the experimental values have depended on time. The calculation of the radiative corrections again provides a lesson in how careful one has to be when dealing with radiative corrections in bound states; in this case, the relativistic corrections to the lowest order graph (fig. 11a) contribute to the same order as the radiative corrections, one of which is shown in fig. 11b. That is, the fact that the electron and positron are in motion relative to each other is just as important as the radiative corrections to the free annihilation cross section in the limit of zero kinetic energy. In this particular case, the motion correction compensates the contribution of the Coulomb photons in fig. 11b, which should be subtracted since it is already included in the bound state wave function. The correct procedure for doing this was given in ref. 31. Further contributions of relative order $\alpha^2 \ln \alpha^{-1}$ were given in ref. 28.

A final remark on positronium deals with the fact that it provides a useful test of one of the fundamental assumptions underlying QED, namely charge conjugation invariance. The decays

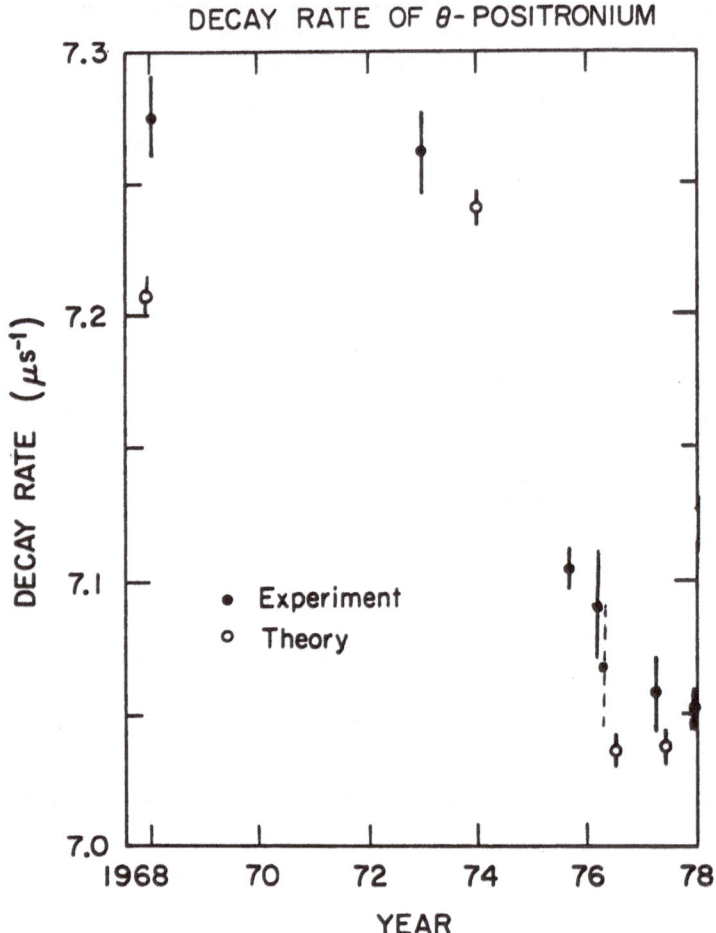

10. Time dependence of theoretical and experimental values for the decay
    rate of orthopositronium.

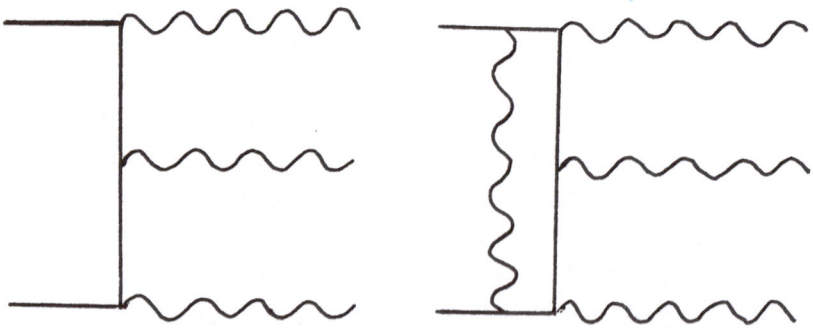

11. (a) Lowest order graph for the decay of positronium to three photons.
    (b) The radiative correction contribution which must be carefully
        treated with regard to contributions due to Coulomb photons and
        binding corrections.

$$(e^+e^-)_{1_{S_0}} \rightarrow 3\gamma \qquad \text{and} \qquad (e^+e^-)_{3_{S_1}} \rightarrow 4\gamma$$

are forbidden by C-invariance. The present upper limits on their branching ratios (relative to the respective allowed process) is 10^{-5} in both cases.[32,33] It would be worth trying to improve on these limits, at least if this can be done as a by-product of other experiments on the positronium lifetimes.

Turning now to atoms having a normal nucleus, one almost always starts with the hydrogen spectrum, which is shown in fig. 12. From the standpoint of tests of radiative corrections, the most interesting feature is the splitting between the $2s_{1/2}$ and $2p_{1/2}$ levels, which would be zero in the absence of all corrections due to radiative effects, relativistic recoil or nuclear extension. The latter two contribute only about 0.5 MHz to the known splitting of 1057.9 MHz; the pure QED contributions dominate the Lamb shift completely.

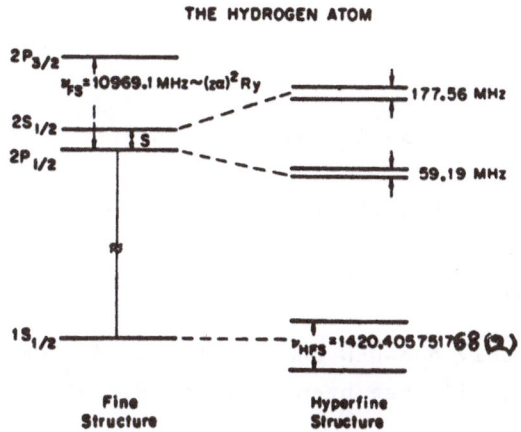

12. Hydrogen spectrum.

At the moment the two most recent published experimental values of the Lamb shift are

$$S_{exp} = 1057.893 \pm 0.020 \text{ MHz}^{34)}$$
$$1057.862 \pm 0.020 \text{ MHz}^{35)}$$

Although these results are consistent within the errors, they are not really in agreement either: it would be nice if one could understand the difference. (Note added in proof: The authors of ref. 34 have recently reported a new value $S_{exp} = 1057.845 \pm 0.009$ MHz.)

The theoretical uncertainties are probably comparable to the experimental ones. In order to see this, it will be necessary to write down the full expression for the Lamb shift*:

$$S_{th} = \frac{\alpha^3 Z^4 R_\infty}{3\pi} \left\{ \left[\frac{19}{30} + \ln\frac{m_e(Z\alpha)^{-2}}{m_r} + \ln\frac{K(2,1)}{K(2,0)} \right] \left(\frac{m_r}{m_e} \right)^3 \frac{<\rho>}{|\phi_{2s}(0)|^2} \right.$$

$$+ \frac{1}{8} \frac{m_r}{m_e}^2 + 0.32206 \frac{\alpha}{\pi} \frac{m_r}{m_e}^3 + 2.2962 \, \pi\alpha Z$$

$$\left. + (\alpha Z)^2 \left[-\frac{3}{4} \ln^2(Z\alpha)^{-2} + 3.9184 \ln(Z\alpha)^{-2} + G_{VP}(Z\alpha) + G_{SE}(Z\alpha) \right] \right\}$$

$$+ S_{size} + S_{rec} + S_{pol} \; .$$

Here

$$S_{rec} = 0.359 \pm 0.011 \text{ MHz}$$

$$S_{size} = 0.1954 <r^2> = 0.144 \pm 0.004 \text{ MHz}^{37}$$

$$S_{pol} \simeq 0.001 \text{ MHz}$$

are the only non-QED contributions. A model-independent calculation of $<\rho>/|\phi_{2s}(0)|^2$ has been given by Borie[36]. The result is

$$<\rho>/|\phi_{2s}(0)|^2 \simeq 1 - 2m_r \alpha Z <r>_{(2)}$$

where $<r>_{(2)} = \int d^3r \int d^3u \; r \, \rho(|\vec{r}-\vec{u}|) \, \rho(u) \simeq 1.05 \pm 0.05$ fm .

---

* The notation of Mohr[10] is adopted. However the dependence on mass and on nuclear size is explicitly included.

The correction term (proportional to $m_r \alpha Z <r>_{(2)}$) gives a previously uncalculated contribution of $- 0.041 \pm 0.003$ MHz.

The value of $G_{SE}(Z\alpha)$ has not yet been unambiguously determined. Two calculations exist, but disagree by more than the experimental uncertainties, also for $Z \neq 1$. Mohr[10] extrapolates numerical values calculated for $Z = 10-30$ using methods developed in ref. 38 while Erickson's expansion[9] was designed for small values of Z. The methods given by Mohr[38] and others[39] have the advantage (in principle) of a-voiding perturbation expansions in αZ with all the previously mentioned pitfalls, but they involve difficult numerical problems which have not yet really been solved for low Z systems. Until the question of the value of $G_{SE}(Z\alpha)$ has been clarified, it is only possible to quote theoretical values for both possibilities:

	$G_{SE}(\alpha)$	contribution to S	S_{th} (MHz)
Erickson	-17.1 ± 1.2	-0.124 ± 0.009	1057.888 ± 0.017
Mohr	-23.4 ± 1.2	-0.169 ± 0.009	1057.843 ± 0.017

Neither possibility is excluded by experiment. Aside from the problem of understanding the higher order binding corrections to the self-energy, also for $Z \neq 1$, theory improvements which should be under-taken will involve higher order recoil corrections (to the non-Breit terms) and corrections to the Bethe sums due to finite nuclear size. The uncertainties in the nuclear parameters will limit the Lamb shift as a test of QED to a few ppm. In contrast to some other tests, however, the Lamb shift is almost entirely a QED effect and thus a more sensitive probe than, for example, the ground state hyperfine splitting of any system.

Muonic atoms provide a complementary test of QED in atoms because the energy levels are sensitive mainly to vacuum polarization, as was discussed previously. In fact the fourth order contribution is nu-merically more important than the muon self energy; thus muonic atoms can provide a test of vacuum polarization which is not significantly disturbed by other QED effects. Of course, in order to minimize per-turbations from other sources, such as nuclear size and structure, or electron screening, it is necessary to choose transitions among orbits such that

$$R_{Nucleus} << R_{Bohr} (\mu) << R_{Bohr} (electrons) .$$

Since the subject of QED tests in muonic atoms has been nicely reviewed elsewhere[15,40] I shall confine my remarks to a few highlights.

The measurements of the Lamb shift and fine structure in muonic helium[41] provide a beautiful example of a precision measurement using laser spectroscopy. Unfortunately, the comparison with theory[42] is limited by the accuracy with which the charge radius of helium has been measured from electron scattering[43] ($<r^2>^{1/2}$ = 1.674 ± 0.012 fm) ,so that the experiment can almost be regarded as a precision measurement of the charge radius of ^{4}He, giving $<r^2>^{1/2}$ = 1.673 ± 0.001 fm .Nevertheless, the accuracy is sufficient to set limits on possible anomalous muon-nucleus interactions, such as might arise from the exchange of a light Higgs boson[40,44].

Table 2:   Comparison between experiment[41] and theory[42] for the 2s-2p transition energies (in meV) for muonic ^{4}He .

	$2s_{1/2} - 2p_{1/2}$	$2s_{1/2} - 2p_{3/2}$
Experiment	1.3813 ± 0.0005	1.5275 ± 0.0003
Theory	1.3809 ± 0.0042	1.5272 ± 0.0042

Recently a proposal has been made to avoid this difficulty by making precision measurements of the $3d_{3/2} - 3p_{3/2}$ and $3d_{5/2} - 3p_{3/2}$ transition energies in muonic helium. By making measurements on states having high orbital angular momentum, for which nuclear effects are much smaller than for s-states, it is hoped that the vacuum polarization calculation can be tested to substantially higher accuracy. Theoretical predictions have been given by Borie and Rinker[46]; for example, vacuum polarization contributes 111.48 meV to the $3d_{3/2} - 3p_{3/2}$ transition energy of 111.42 meV.

Tests of QED in heavy muonic atoms can test the higher order vacuum polarization contributions, particularly those arising from Coulomb corrections to the electron propagator. Since the subject is reviewed here[15] and elsewhere[40] I shall merely remark that earlier indications of a discrepancy between theory and experiment (see for example ref. 47) have been eliminated by more recent experiments and more accurate measurements of calibration lines. Summarizing, one can say that vacuum polarization has been tested at the level of about 0.2 % (2000 ppm) while the higher order corrections have been tested to about 15-20 % over a wide range of atomic numbers from Z = 2 to Z = 82.

In conclusion, there appear to be no major discrepancies between theory and experiment in atomic physics tests of QED, although of course this comes as no surprise. The theoretical calculations need improvement, not only for the case of positronium and muonium, where the need has been recognized for some time, but also for hydrogenlike atoms, where it would be helpful to have a better understanding of higher order binding, relativistic recoil, and finite size corrections. Experimental input is also needed, mainly with regard to more accurate measurements of the parameters of the theory.

The author wishes to thank Dr. H. Pilkuhn for several helpful suggestions and discussions during the preparation of this talk.

References:

1. For more comprehensive reviews see
    B.E. Lautrup, A. Peterman, E. de Rafael, Phys. Rep. 3C (1972) 193
    S.D. Drell, Physics 96A (1979) 3
2. G.P. Lepage, Phys. Rev. A16 (1977) 863
3. W.E. Caswell, G.P. Lepage, Phys. Rev. A18 (1978) 810
4. G.T. Bodwin, D.R. Yennie, Phys. Rep. 43C (1978) 267
5. R. Barbieri, E. Remiddi, Nucl. Phys. B141 (1978) 413
6. H. Grotch, D.R. Yennie, Rev. Mod. Phys. 41 (1969) 350
7. H. Pilkuhn, Relativistic Particle Physics, Springer Verlag,
    New York (1979) p. 71
8. J.M. Jauch, F. Rohrlich, The Theory of Photons and Electrons,
    2nd Ed., Springer Verlag, New York (1976) Sect. 15.4
9. G.W. Erickson, J.Phys. Chem Ref. Data 6 (1977) 831
10. F.J. Mohr, Phys. Rev. Lett. 34 (1975) 1051
11. E.A. Ühling, Phys. Rev. 48 (1935) 55
12. L.W. Fullerton, G.A. Rinker, Phys. Rev. A13 (1976) 1283
13. E.H. Wichmann, N.M. Kroll, Phys. Rev. 101 (1956) 843
    G.A. Rinker, L. Wilets, Phys. Rev. A12 (1975) 748
    M. Gyulassy, Nucl. Phys. A244 (1975) 497
14. G. Källén, A. Sabry, K. Dan. Vidensk. Selsk Mat.-Fys. Medd. 29,
    No. 17 (1955)
15. L. Tauscher, this conference
16. E. Borie, J. Calmet, this conference
    H.G. Dehmelt, this conference
17. S.J. Brodsky, G.W. Erickson, Phys. Rev. 148 (1966) 26
18. D.E. Casperson et al., Phys. Rev. Lett. 38 (1977) 956
19. W.E. Caswell, G.P. Lepage, Phys. Rev. Lett. 41 (1978) 1092
20. E.R. Williams, P.T. Olsen, Phys. Rev. Lett. 42 (1979) 1575
21. H. Orth, this conference
22. E. Borie, Z. Phys. A291 (1979) 107
23. E.W. Weber, this conference
24. P.O. Egan et al., Phys. Lett. A54 (1975) 412
25. A.P. Mills, G.H. Bearman, Phys. Rev. Lett. 34 (1975) 246
26. T. Fulton, Phys. Rev. A7 (1973) 377
27. V.K. Cung, A. DeVoto, T. Fulton, W. Repko, Nuovo Cim. 43A (1978)
    634; Phys. Rev. A19 (1979) 1886
28. W.E. Caswell, G.P. Lepage, Phys. Rev. A20 (1979) 36
29. M. Stroscio, Phys. Rep. 22C (1975) 217
30. P. Zitzewitz, this conference

31. W.E. Caswell, G.P. Lepage, J. Spirstein, Phys. Rev. Lett. <u>38</u> (1977) 488

32. A.P. Mills, S. Berko, Phys. Rev. Lett. <u>18</u> (1967) 420

33. K. Marko, A. Rich, Phys. Rev. Lett. <u>33</u> (1974) 980

34. S.R. Lundeen, F.M. Pipkin, Phys. Rev. Lett. <u>34</u> (1975) 1368

35. D.A. Andrews, G. Newton, Phys. Rev. Lett. <u>37</u> (1976) 1259

36. E. Borie, TKP 80-8, to be published

37. The proton rms charge radius was taken from G.G. Simon et al., Z. Naturforsch. <u>35a</u> (1980) 1

38. P.J. Mohr, Ann. Phys. (N.Y.) <u>88</u> (1974) 26, 52

39. A.M. Desiderio, W.R. Johnson, Phys. Rev. <u>A3</u> (1971) 1267
    K.T. Cheng, W.R. Johnson, Phys. Rev. A14 (1976) 1943

40. E. Borie, Habilitationsschrift, University of Karlsruhe, 1979
    E. Borie, G.A. Rinker, to be published

41. G. Carboni et al., Nucl. Phys. <u>A278</u> (1977) 381; Phys. Lett. <u>73B</u> (1978) 229

42. E. Borie, G.A. Rinker, Phys. Rev. <u>A18</u> (1978) 324

43. I. Sick, J.S. McCarthy, R.R. Whitney, Phys. Lett. <u>64B</u> (1976) 33

44. H.J. Leisi, Nucl. Phys. <u>A335</u> (1980) 3

45. E. Zavattini, A.M. Sachs, J. Fox, R. Cohen, AGS proposal, April 1979

46. E. Borie, G.A. Rinker, Z. Phys. A, in press

47. R. Engfer, J.L. Vuilleumier, E. Borie, in Atomic Physics <u>4</u>, ed. by G. zu Putlitz, E.W. Weber, A. Winnacker, Plenum Publ. Co. (1975) p. 191

GROUND STATE POSITRONIUM: HYPERFINE STRUCTURE AND DECAY RATES

P.W. Zitzewitz

Department of Natural Sciences
University of Michigan-Dearborn
Dearborn, MI 48128 USA

(1979-80 Fakultät für Physik
Universität Bielefeld
4800 Bielefeld 1 West Germany)

To be published in the Proceedings of the Symposium on the Status and
Goals of Quantum Electrodynamics, Mainz, W. Germany, May 9-10, 1980.

A. INTRODUCTION

Positronium, the bound state of an electron and its antiparticle, the positron, is an attractive testing ground for quantum electrodynamics because its constituent particles interact to high order only through the electromagnetic interaction. The existence of positronium (Ps) was suggested by Mohorovicic in 1934 and many of its properties were calculated by Pirenne in 1943.[1]

The gross (Bohr) energy levels are hydrogenic, but with half the energy because the reduced mass is half the electron mass. The ground state, as in hydrogen, is composed of three triplet and one singlet spin states. The spin-spin hyperfine splitting is much larger than in hydrogen (about 200 GHz rather than 1.4 GHz), partly because of the large positron magnetic moment, but also because of a QED effect, virtual Ps creation and annihilation. The triplet state decays only (by charge conjugation invariance) into an odd number (greater than one) gammas, three primarily, with an average lifetime of 142 nsec. The singlet decays into an even number, principally two, in about 0.1 nsec. In 1952 Deutsch reported the production of positronium and the first crude measurements of its hyperfine separation and triplet lifetime.

The effect of an external magnetic field on the ground state is an interesting quantum mechanical effect, crucial to most experimental techniques. The $m = \pm 1$ components of the triplet (represented by $\downarrow\uparrow$ and $\uparrow\downarrow$ where $\downarrow$ is the positron spin direction and $\uparrow$ the electron) have a net magnetic moment of zero, and thus are unaffected by an external field.

However, the m = O levels, $\frac{1}{\sqrt{2}}$ (↓↑+↑↓) for the triplet and

$\frac{1}{\sqrt{2}}$ (↓↑-↑↓) for the singlet at zero magnetic field, are

no longer eigenstates in an external field. This may be seen

by considering an extremely strong external field, in which

↑↓ and ↓↑ are eigenstates. Each of these is a 50%-50% super-

position of the zero-field singlet and triplet states. Thus

in intermediate fields the perturbed triplet contains a field-

dependent singlet component. Likewise, the perturbed singlet

contains some triplet admixture. Correspondingly, the decay

modes are partially via three and partially via two gammas,

and the lifetimes are intermediate. For example, in a 5kG

field the perturbed triplet decays via 99.5% 3-γ, 0.5% 2γ

events with a mean lifetime of 22.5 nsec.

B.   HYPERFINE SEPARATION

  i) Theoretical

     The theoretical calculation of the hyperfine separation

in Ps is made difficult by the lack of a heavy nucleus, which

prevents reduction to a one-body problem. Traditionally use

has been made of the Bethe-Salpeter equation, the two-body

formulation of relativistic quantum mechanics, which has no

unique way to reach an approximate solution. Recently,

Lepage, et al.,[2] have developed an equivalent Schroedinger

equation with reduced mass, an entirely different approach.

The results of the calculations, listed in Table I, are conveniently organized by order in α. The zeroth order term was calculated by Pirenne, the first order by Karplus and Klein in 1952. The second order terms include those with $\alpha^2 \ln \alpha^{-1}$ as well as α^2. The former have proved particularly troublesome because from time to time new contributions are found, and the size of the terms is large because $\ln \alpha^{-1} = 4.9$. The latter are only partially calculated. So far the static, recoil, vacuum polarization, and two and three photon virtual annihilation terms have been completed.

The present experimental results, discussed below, are at the 6ppm level. Thus completion of the second order calculations will allow testing the coefficients of $\alpha^2 \ln \alpha^{-1}$ to 5% and those of α^2 to 25%, since the next order terms should contribute only about 1 ppm.

ii) Experimental

All experiments involving Ps are hampered by the difficulty in forming reasonably large samples. Positrons come from radioactive sources with energies in the 200 keV range and must be slowed to about 10 eV in order to capture an electron from an atom and form Ps. The source of atoms is traditionally the material in which the positron energies are moderated, usually a gas. Collisions with these atoms can then perturb the properties of the Ps.

T A B L E   I

Positronium Hyperfine Separation
Theory

Order	Source	Contribution in units α^2 Ry	in GHz
α^4	Magnetic plus Virtual Annihilation	$\dfrac{7}{6}$	204.3867
α^5	Bethe-Salpeter Radiative Corrections	$-\dfrac{\alpha}{\pi}\left(\dfrac{16}{9} + \ln 2\right)$	$-$ 1.0055
$\alpha^6 \ln \alpha^{-1}$	Radiative Corrections	$+\dfrac{5}{12}\,\alpha^2 \ln \alpha^{-1}$	$+$ 0.0191
α^6	Virtual Annihilation Three Loop Radiative Corrections	$-0.32\,\alpha^2$	$-$ 0.0045
	Total		203.3958 GHz

The hyperfine transition is at high frequency (203 GHz) and requires so much power that, even today, direct excitation is unfeasible. Rather, the Ps is formed in a magnetic field ($\approx$ 8 kG) and the Zeeman (triplet m = ±1 to m = 0) transition is driven.[3] The perturbation of the m = 0 state by the field, as previously explained, means that a fraction of the decays from that state are via two gammas rather than three. Thus a measurement of the $2\gamma/3\gamma$ ratio will show a small ($\approx$ 6%) increase at resonance. Because high power is needed to drive the transition a microwave cavity is required and used as the gas container (see Fig. 1), and the magnetic field, H, rather than the frequency, is swept. The hyperfine frequency, $\Delta\nu$, is found by measuring the Zeeman frequency, f_{o1}, and solving

$$f_{o1} = \tfrac{1}{2} \Delta\nu \; [(1+X^2)^{\tfrac{1}{2}}-1], \text{ where } X = 2g'\mu_B H/h\Delta\nu,$$

and g' is the g-factor of the bound lepton, shifted by 11.1 ppm from the free-space value. A large number of data points, typically 10^{10}, must be collected to obtain the needed statistical precision in view of the small signal.

The natural linewidth, determined by the short lifetime of the singlet decay, is large, $\approx$ 0.6% of the transition frequency. To obtain the 6 ppm accuracy of the present experiments, the line must be split to a part in a thousand. Thus, careful consideration must be given, from both the theoretical and experimental side, to the resonance line shape. A further systematic effect must be considered;

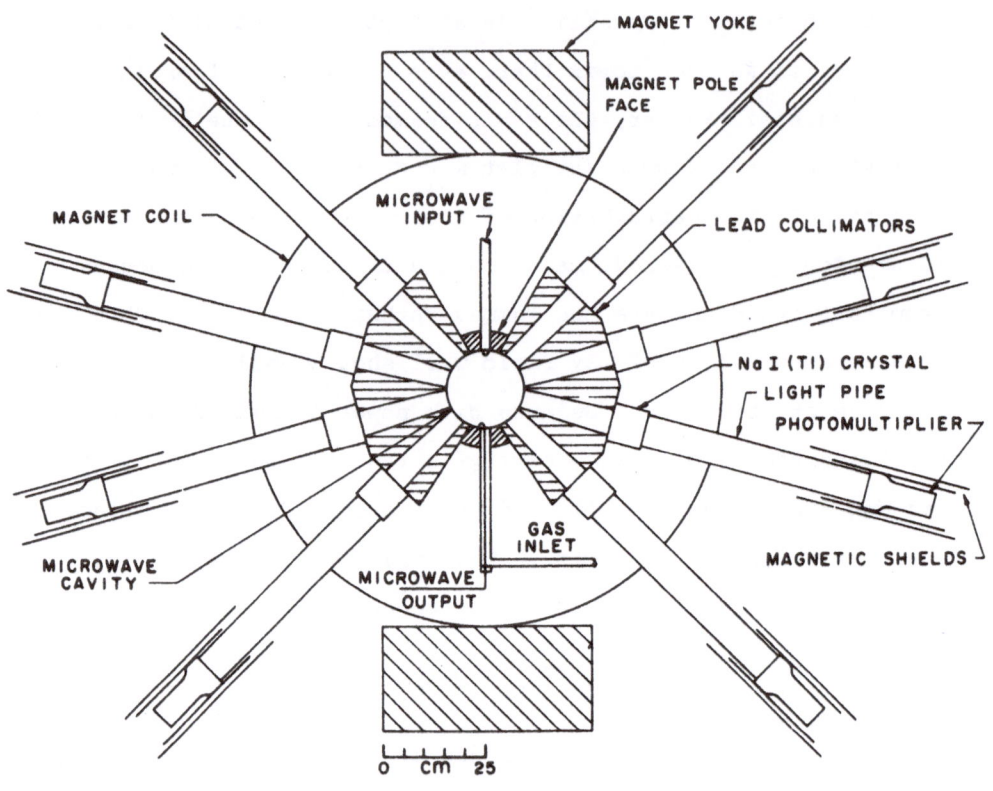

Fig. 1

Positronium hyperfine separation apparatus at Yale. (Ref.3)

the shift of the hfs by collisions of Ps with the buffer gas. The shift is corrected by measuring the frequency as a function of gas density and extrapolating to zero. N_2 and SF_6 were used in the two experiments, but shifts of all the noble gases have been measured.

The two most recent results are $\Delta\nu = (203.3849\pm0.0012)$ GHz from the Yale group,[3] and $\Delta\nu = (203.3870\pm0.0016)$ GHz from the Brandeis group.[4] They are respectively 10 and 6 standard deviations from the theoretical value of 203.3958 GHz. Whether or not this is a real disagreement awaits the completion of the order (α^2) contributions. Further improvements in the experimental system are possible. By increasing the radio-active source strength and the number of detectors, as well as improving the accuracy in determining the magnetic field, a level of 2-3 ppm could be reached.

C. LIFETIMES

i) Theory

The lowest order calculation of the decay of singlet (para) positronium (p-Ps) into two gammas was made by Pirenne in his classic paper. In 1949 Ore and Powell calculated the corresponding contribution to the decay rate of triplet (ortho) positronium (o-Ps) into three gammas. In 1952-3 Wolfenstein and Ravenhall showed that charge conjugation invariance demands that o-Ps decay into an odd number of gammas (greater than one) while p-Ps decay into an even number.

In 1957 Harris and Brown succeeded in calculating the radiative corrections to the decay rate of p-Ps. These were, in fact, the first calculations of radiative corrections to any decay rate. As can be seen in Table II, these can also be expressed in terms of orders of α. The order α term is small, about 0.5%, and has been confirmed recently by Cung[5] and by Freeling.[6] Caswell and Lepage[7] as well as Tomozawa[8] have confirmed the result and extended it by calculating the $\alpha^2 \ln \alpha^{-1}$ term.

The radiative correction to the o-Ps decay rate was first calculated 17 years later. Stroscio and Holt reported a result in 1974 which was revised a year later by Stroscio to $(1.86\pm0.45) \frac{\alpha}{\pi} \lambda_0$,[9] where the error comes primarily from the uncertainty in calculating the numerical integrals. The resulting decay rate, $\lambda = (7.242\pm0.008)$ $\mu\mathrm{sec}^{-1}$ was in good agreement with the two experimental results then existing, $\lambda = (7.262\pm0.015)$ $\mu\mathrm{sec}^{-1}$ from the London group[10] and $\lambda = (7.275\pm0.015)$ $\mu\mathrm{sec}^{-1}$ from the Yale group.[11] As will be discussed shortly, however, it was soon in strong disagreement with the results obtained in Michigan in 1975-77. This disagreement prompted the SLAC group to use their techniques developed for the hyperfine separation to calculate the o-Ps decay rate. They found an error in the sign of one term and improved the numerical calculation of others, yielding a final result of $\lambda = (7.0386\pm0.0002)$ $\mu\mathrm{sec}^{-1}$.[2] This is a

T A B L E   II

Positronium Decay Rates - Theory

Parapositronium   (p-Ps) $1\ ^1S_o$

Order	Rate
$\Gamma^o = \dfrac{\alpha^5 mc^2}{2h}$	$8.03265\ \text{nsec}^{-1}$
$\Delta\Gamma^1 = \Gamma^o(-\dfrac{\alpha}{\pi})\ (5 - \dfrac{\pi^2}{4})$	$-\ 5.88\cdot10^{-3}\ \Gamma^o$

$$\text{Total}\quad \Gamma_p = 7.9854\ \text{nsec}^{-1}$$

Orthopositronium   (o - Ps) $1\ ^3S_1$

$\Gamma^o = \dfrac{2}{9}\ \dfrac{\alpha^6 mc^2}{h}\ (\dfrac{\pi^2-9}{\pi})$	$7.2112\ \mu\text{sec}^{-1}$
$\Delta\Gamma^1 = \Gamma^o\ (-10.266\pm0.011)\ \dfrac{\alpha}{\pi}$	$-\ 0.1720\pm0.0002\ \mu\text{sec}^{-1}$
$\Delta\Gamma^{2'} = \Gamma^o\ (-\dfrac{1}{3})\ \alpha^2\ell n\alpha^{-1}$	$-\ 0.0006\ \mu\text{sec}^{-1}$
$0(\Delta\Gamma^2) = 0(\dfrac{\alpha}{\pi})^2\ \Gamma^o$	$\approx\ \pm0.0004\ \mu\text{sec}^{-1}$

$$\text{Total}\ \Gamma_o = (7.0386\pm0.0002)\ \mu\text{sec}^{-1}$$

change in the value on the order of 3% ! The $\left(\frac{\alpha}{\pi}\right)^2$ terms are not yet calculated, and could contribute (0.0004) μsec^{-1} if the coefficient has a magnitude of 10, the size of the coefficient of (α/π).

ii) Experiment

The only experimental measurement of the decay rate of p-Ps is an indirect measurement which used the linewidth in the hyperfine separation experiment.[12] While the result, $\lambda = (7.99\pm0.11)$ $nsec^{-1}$, is in agreement with theory, the uncertainty is much larger than the entire radiative correction. As technical limitations dictated by the extremely short decay time prohibit a direct measurement, a second, completely different, indirect measurement is now underway at Michigan. It is based on a measurement of the decay rate of the magnetic-field perturbed triplet state. This decay rate λ is given by a linear combination of the unperturbed triplet and singlet rates, that is

$$\lambda = \frac{1}{1+y^2} [\lambda_t + y^2 \lambda_s]$$

where $y = x/(1 + x^2)$ and $x = 0.0276$ B, where B is in kilogauss. A simultaneous measurement of the unperturbed and perturbed triplet rates can be used to give the singlet decay rate. Systematic effects of collisions with the buffer gas in a magnetic field will have to be eliminated in order to obtain the 0.2% precision in λ_s which appears feasible. This should

be sufficient to verify the first order radiative correction
term. A similar experiment, but using a low energy positron
beam and no buffer gas is underway at Mainz.

In contradistinction to the p-Ps measurements, all
precision measurements of the o-Ps decay rate have used a
direct method. Before reviewing the results, it is worthwhile
outlining the methods used. Two timing signals are obtained,
one from the birth of positronium, one from the annihilation
(see Fig. 2). The pulse from the annihilation is obtained
by the use of a photomultiplier on a gamma-ray sensitive
scintillator. The pulse from the formation is obtained
indirectly from either the gamma-ray which is emitted in
coincidence with the emission of the positron from ^{22}Na or
from the passage of the positron through a thin plastic scintillator. The time required for positron energy loss and Ps
formation is very short ($\approx 10^{-11}$ s) and can be found from the
signal from direct annihilation of the positrons together with
decays from p-Ps (the "prompt" signal). The pulses are shaped
with fast discriminators. In some experiments a rejection
system is used to eliminate events for which the start and stop
signals are within 25 nsec of each other, and thus reduce the
background. The time interval between the two pulses is converted
into a pulse whose amplitude is proportional to the time interval
and the spectrum of these pulse amplitudes is recorded on a multichannel analyzer. Alternatively, for some experiments the
time interval can be analyzed and converted into a digital
signal which is transferred directly into the memory of the MCA.

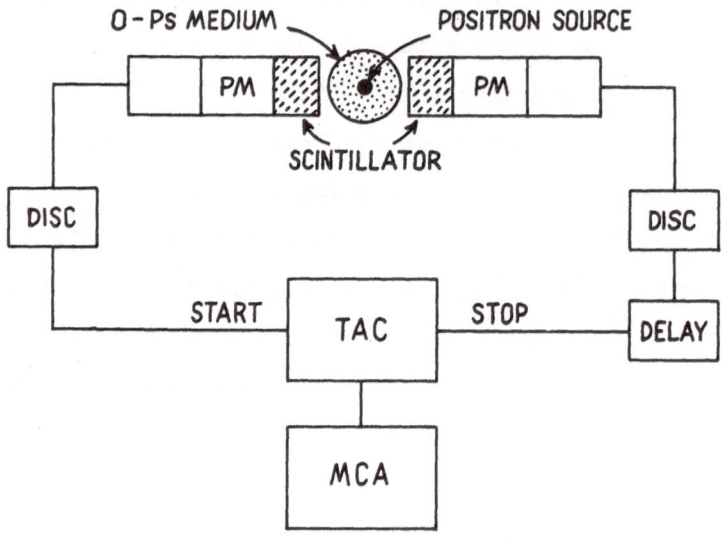

Fig. 2

General schematic diagram of the method of directly
measuring the o-Ps decay rate.

As in the hyperfine measurements, the high energy positrons must be moderated to about 10 eV in order to bind an electron from an atom and from Ps. In the first two measurements, the results of which are quoted above, gases were used. However, in 1968 it was shown in Paris[12] that certain metal oxide powders with extremely small particle size copiously form positronium with a lifetime near the free-space value. Since the lifetime of Ps within matter is under 2 nsec, it was proposed that the Ps was ejected from the particle and existed in the intergrain region. The first precision decay rate measurements in this new medium were performed at Michigan in 1975 (see Fig. 3) with the surprising result that some measurements in powder (as shown in Fig. 4) gave a lifetime <u>longer</u> than the previous extrapolation to free space[13]. It was also shown that the decay rate depended linearly on the free volume density between the powder grains (see Fig. 4), where the positronium is likely to exist. After careful examination of the experiment, as well as consideration of possible effects of the powder surfaces which would not extrapolate to zero, Ford, Sander, and Witten[14] concluded that the extrapolated value $\lambda = (7.104 \pm 0.006)$ μsec^{-1} probably did represent the free-space value.

In order to provide a completely independent check of this result, a second measurement was begun at Michigan which used another technological innovation. It had been found in 1958 that when high energy positrons are incident on solids a small fraction (from 10^{-7} to 10^{-3}) are re-emitted with energies below 1 eV[15]. These could be formed into a well

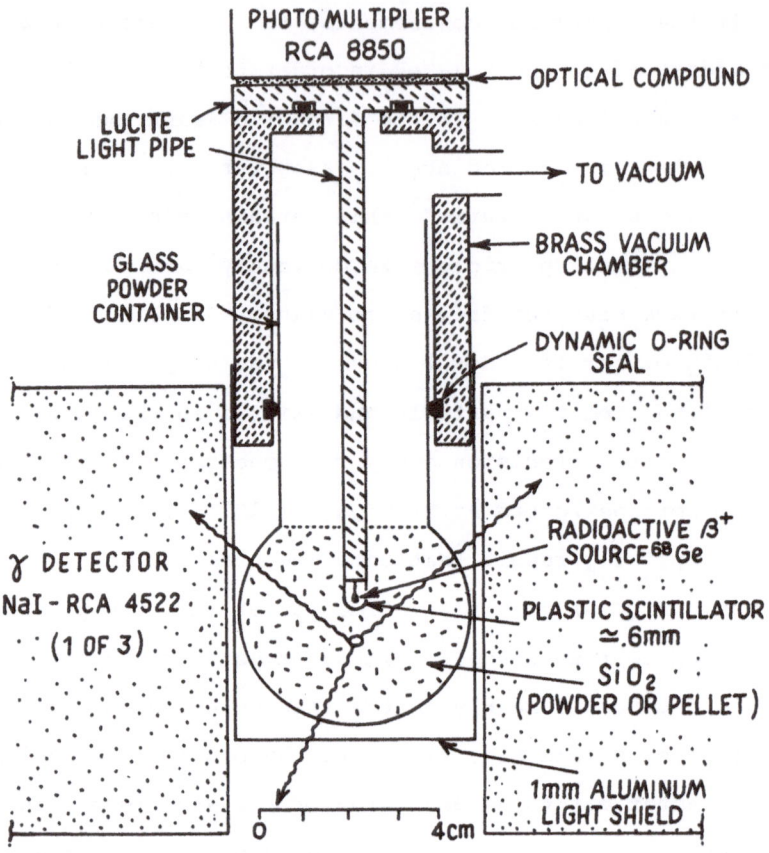

PHOTO MULTIPLIER RCA 8850

OPTICAL COMPOUND

LUCITE LIGHT PIPE

TO VACUUM

GLASS POWDER CONTAINER

BRASS VACUUM CHAMBER

DYNAMIC O-RING SEAL

γ DETECTOR NaI - RCA 4522 (1 OF 3)

RADIOACTIVE β^+ SOURCE ^{68}Ge

PLASTIC SCINTILLATOR $\simeq$.6mm

SiO_2 (POWDER OR PELLET)

1mm ALUMINUM LIGHT SHIELD

0   4cm

Fig. 3

Apparatus used to measure the o-Ps decay rate in powders at Michigan in 1974.

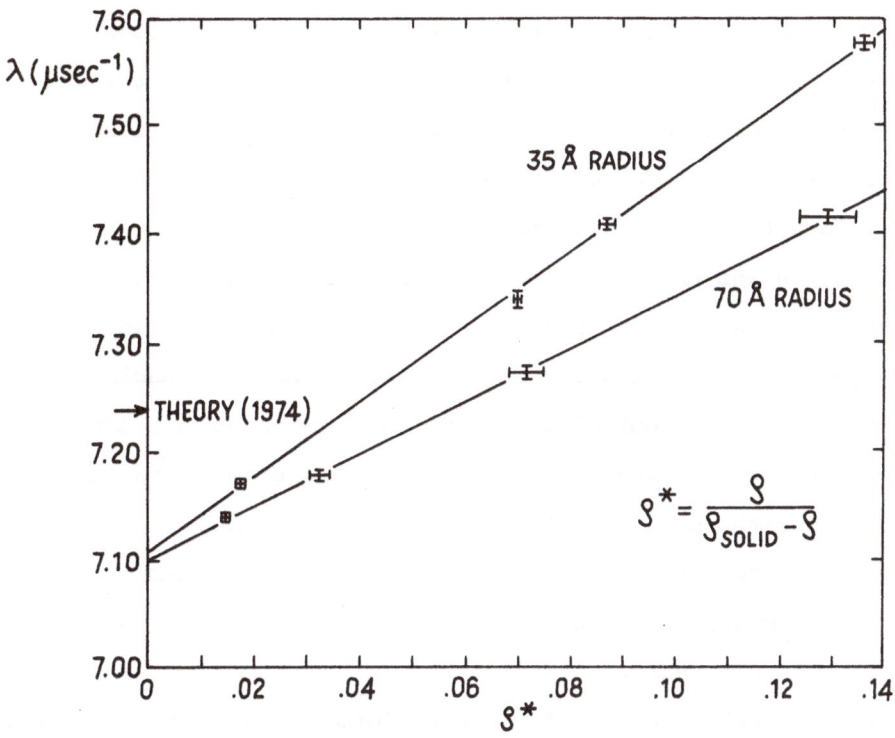

Fig. 4
Decay rate as function of reduced powder density.

focussed, essentially monoenergetic beam with adjustable
energy (see Fig.5). It was shown at Brandeis[16] that when
such a beam of energy ~ 400 eV was incident on metal surfaces
positronium was formed and ejected into vacuum. The
surface for positronium formation used at
Michigan was the secondary-electron emitting surface of a
channel-electron multiplier (CEM). This also provided a
signal for the formation of the positronium. By enclosing
the region in front of the CEM (see Fig.6) with a metal
can suitably coated so that the positronium atoms would
bounce with very low probability of annihilation, it was
possible to make a direct measurement of the vacuum decay
rate. To support this claim it was shown that the Ps actually
left the CEM surface. This was done by placing a gamma
detector behind a slot in lead to form a gamma-ray telescope.
The telescope accepted only those gammas from positronium
annihilating 5 cm from the CEM. It was found, as shown in
Fig. 7, that the first of these Ps atoms arrived 60 nsec
after the CEM pulse.   An analysis of the shape gave
an average Ps energy of 2.7 eV. A coating of MgO, formed
by exposing the confinement can surface to the smoke from
burning magnesium, was found to be a surface on which the Ps
did not annihilate. Rough measurements showed that at least
80% of the Ps bounced off this surface. Ps atoms could leave
the confinement can through the entrance hole for the
·positrons. The effect of this hole is to give an added

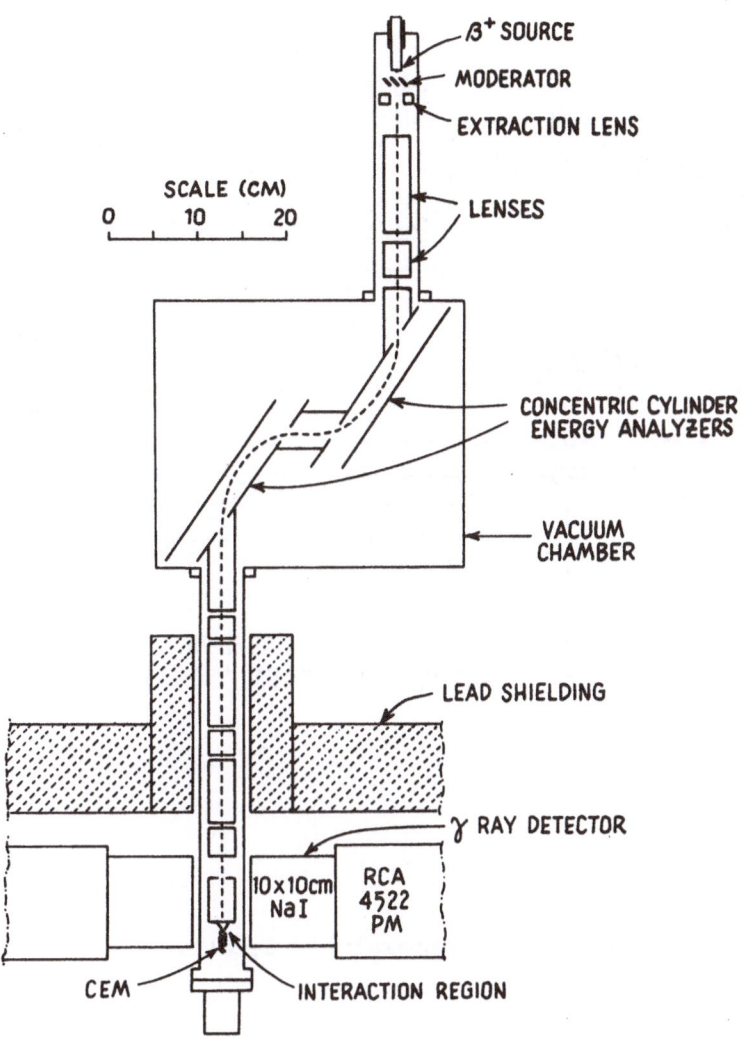

Fig. 5

"Vacuum" decay rate apparatus - slow positron beam,
Michigan, 1975.

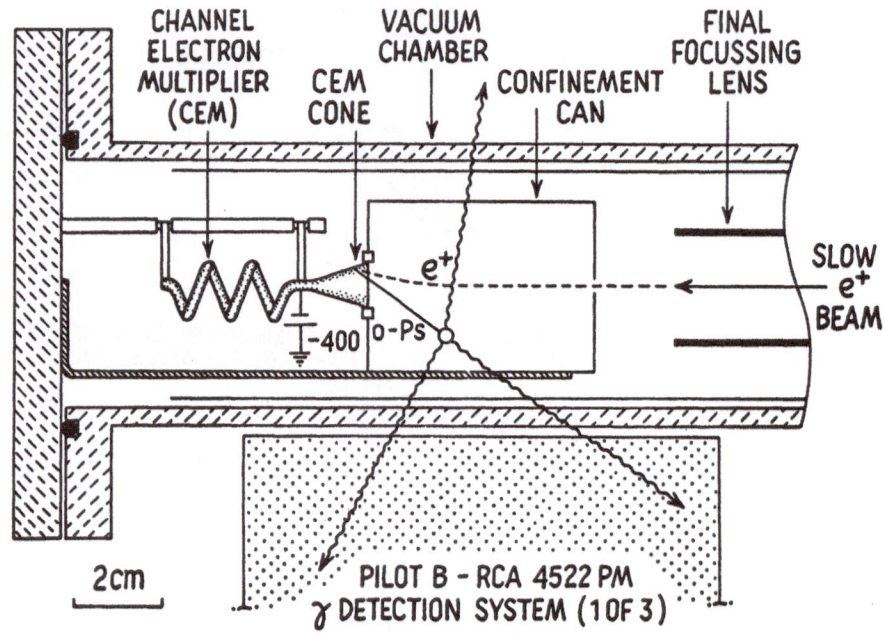

Fig. 6

o-Ps "Vacuum" decay rate apparatus - detail.

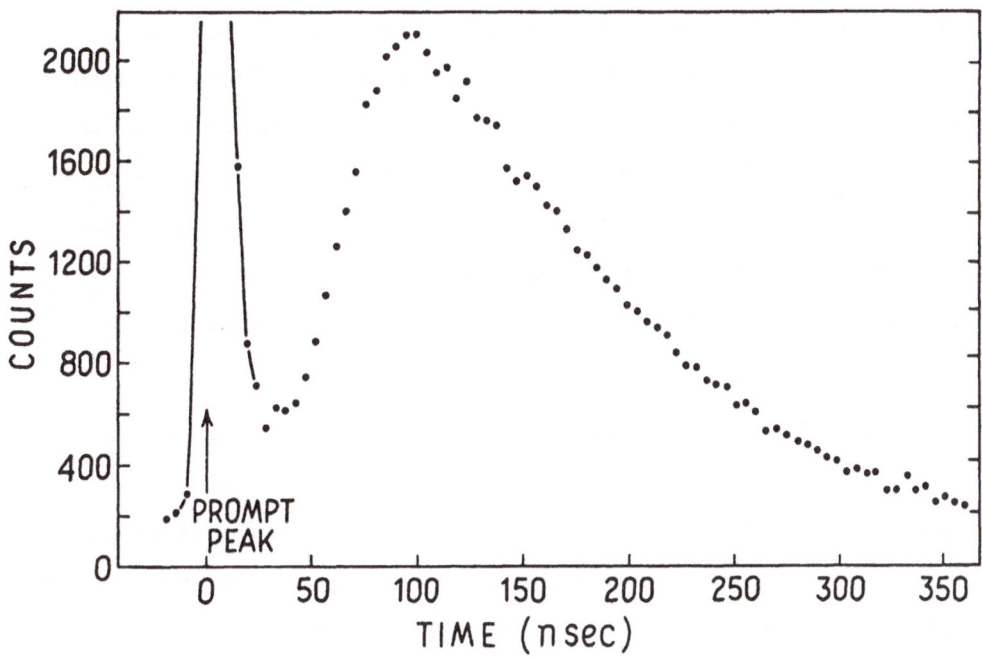

Fig. 7

o-Ps time-of-flight data. Detector is (50±5) mm from CEM.

channel for the disappearance of the Ps, and thus an
increased decay rate. The effect was estimated and the published
result was $\lambda = (7.09\pm0.02)$ μsec^{-1}.[17] The uncertainty was
primarily systematic, caused by an apparent oscillation in the
decay rate found by computer fitting the observed spectrum to
the form Ae$^{-\lambda t}$+B as events at early times are stripped away.
It was the publication of this result, in agreement with the
powder value, but >2% lower than previous experiments and
theory, which prompted the SLAC group to recalculate the
radiative corrections.

As a check a remeasurement of the decay rate using gas
as moderator was made. Since it appeared that problems in
the earlier gas measurements were due to Ps annihilations on
the walls of the confinement vessel at low pressure, the
vessel in this experiment was designed to have a variable
surface/volume ratio. Moreover, by inserting an aluminum
"honey comb" into the chamber, the surface to volume ratio
could be increased by a factor of 20. However, no decrease
in the decay rate was noted at the 0.007 μsec^{-1} level and thus
there exists no known reason for the high decay rate of the
early experiments. In the course of these measurements it was
found that the long light decay curve of NaI(Tl) scintillators
gave rise to low level (10^{-4}) after pulsing in the gamma
detection channel that led to systematic shifts in the fitted
decay rate. Replacing them with Pilot B plastic scintillators
made the fitted decay rate constant at a lower value. A

total of six sets of experiments were made, each set consisting
of measurements at pressures from 200 to 1600 Torr in Freon 12,
isobutane, or a mixture. A number of conditions were varied
to set limits on systematic shifts. The final published result
was λ = (7.056±0.007) μsec^{-1}.[18] Uncertainties are primarily
statistical with some systematic uncertainty from possible
surface annihilation, gas pressure error, and pressure -to-
density conversion.

Discovery of the systematic shift due to the NaI(Tl)
scintillators led to a remeasurement of the decay rate in
powder. A second systematic error was found here, involving
measurement of the powder density, and the new result was
λ = (7.067±0.021) μsec^{-1}.

Finally, a second experiment using the "vacuum"
production was run. Again, Pilot B scintillators were used,
the positron beam strength was increased, and a study of
the effect of the positron entrance hole on the decay rate
was carefully made. Gas kinetic considerations show that
collisions with the can surface and entrance hole should lead
to a measured decay rate λ_m given by

$$\lambda_m = \lambda_o + P_a \frac{v}{4} \frac{S}{V} + cv \frac{A}{V}$$

where S is the area of the can, A the area of the hole, V the
volume of the can, P_a probability of annihilation per
collision, and v the positronium velocity. The results for
one can with fixed S/V is shown in Fig.8. Results from

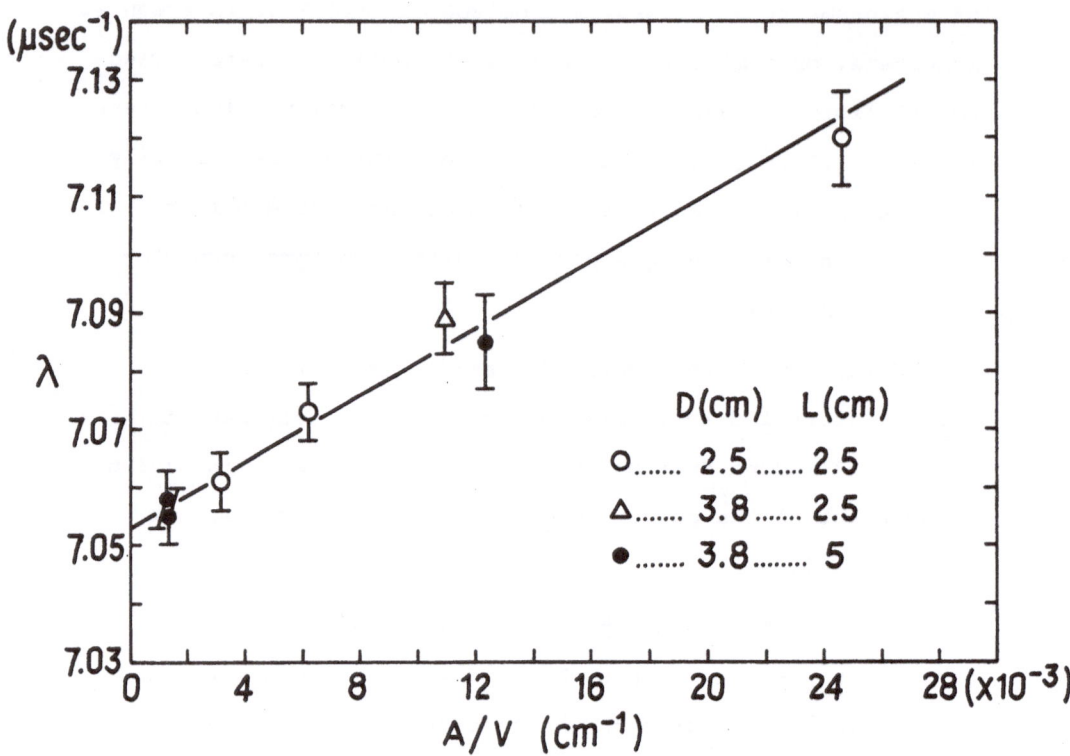

Fig. 8
Decay rate as a function of area of confinement can hole.

different S/V ratio cans show no difference, allowing a upper limit to be placed on $P_a = 10^{-4}$. The final result was $\lambda = (7.050\pm0.013)$ μsec^{-1}.

An independent measurement using the gas method has been made by the University College-London group, who obtained $\lambda = (7.045\pm0.006)$ μsec^{-1}. A weighted average of the four experimental results gives $\lambda = (7.050\pm0.004)$ μsec^{-1}, where the uncertainty is the inverse root of the sum of the weights. The uncertainty probably should be increased to 0.006 μsec^{-1} because the uncertainty in the London result is primarily systematic, while that of the Michigan gas experiment is mostly statistical, and these two contribute most strongly to the average value.

The difference between theory and experiment is then (0.11 ± 0.006) μsec^{-1}. We consider this to be, most likely, a real discrepancy, probably due to an experimental systematic effect. Our conclusion is based on the fact that most of the systematic effects tend to raise rather than lower the measured decay rate.

What improvements can be made in the experiments? Despite the high rate Ps formation, and thus the greater statistical precision in the powder experiments, the determination of powder density and its homogeneity, as well as possible effects which are non-linear in powder density, make improved experiments in this medium the most difficult.

If the problem of assuring that the CEM cone does not have
a higher annihilation rate can be solved, and we believe
it can, then there is no fundamental limitation on the
ability of the vacuum method to yield improved precision.
The present effort at Michigan is in the gas method. About
half the present uncertainty is in statistical and half in
systematic effects. The systematic effects are primarily in the
area of pressure measurement and pressure-density conversion
(gas virial coefficients must be known at the 1% level).
While improvements here will be difficult, they are not
impossible. The statistical proplems are related to the
low formation rate of Ps. An experiment is now underway
which confines the positrons to a small region at the center
of the gas chamber by means of a magnetic field. (The short
diffusion distance of Ps thus insures the it is also close
to the center of the chamber, reducing possible wall effects.)
The longer path length of the positrons leads to a high data
rate with no loss in signal-to-noise. It is hoped that the
resulting reduction in statistical uncertainty to a part in
10^4 can be matched with a corresponding improvement on the
systematic side as just discussed. The completion of these
two experiments, with totally different systematic effects
and with a reduction in uncertainty by a factor of almost ten,
should conclusively show whether or not the discrepancy exists.

D. CONCLUSION

On the theoretical side, effort on the hyperfine structure is most important because of the large size of the coefficients of the higher order terms. On the experimental side, while straightforward improvements in the present techniques are possible, most progress in the past five years has come from the introduction of new techniques, for example, powders and slow positron beams. Especially in view of the overwhelming importance of systematic effects in precision experiments, it is most likely that great improvements will come from presently unknown directions. Finally, looking at the work of the last ten years shows how difficult it is to keep earlier experimental or theoretical results from influencing the result obtained in new work.

E. ACKNOWLEDGEMENTS

The work at Michigan reported here is supported by the Program in Atomic, Molecular, and Plasma Physics of the National Science Foundation. I would like to thank Prof. A. Rich and Dr. D. W. Gidley for careful reading of this manuscript, and Prof. W. Raith for his hospitality, and the Alexander von Humboldt Stiftung for a fellowship during the time in which this manuscript was prepared.

REFERENCES

1   A comprehensive review of positronium research is to be
     published by A. Rich in Rev. Mod. Phys. (1981).
     The most recent published review of fundamental Ps research
     is S. Berko, K.F. Canter, and A.O. Mills, Jr., Progress in
     Atomic Spectroscopy, B, Ed. W. Hanle and H. Kleinpoppen,
     p. 1427, 1979. Basic papers are referenced therein.

2   Caswell, W.E. and G.P. Lepage, Phys. Rev. A20, 36 (1979).

3   Egan, P.O., V.W. Hughes, and M.H. Yam, Phys. Rev. A15,
     251 (1977).

4   Mills, A.P., Jr., and G.H. Bearman, Phys. Rev. Lett.
     34, 246 (1975)

5   Cung, V.K., Devoto, A., Fulton, T., and Repko, W.W.,
     Phys. Lett. 688, 474 (1977).

6   Freeling, J.R., Ph.D.Thesis, Univ. of Michigan, 1979
     (unpublished)

7   Caswell, W.E. and G.P. Lepage, Phys. Rev. A18, 810 (1978)
     and ref. 2

8   Tomozawa, Y.   1979 (to be published)

9   Struscio, M.A., Phys. Rev. A12, 338 (1975)

10   Coleman, P.G. and Griffith, P.G., J. Phys. B 6, 2155 (1973)

11   Beers, R.H., and Hughes, V.W., Bull. Am. Phys. Soc. 13,
     633 (1968), and Hughes, V.W., Plenarvorträge,
     37. Physikertagung der D.P.G., p. 123 (1973).

12   Paulin, R. and Ambrosino, G., J. Phys. 29, 263 (1968)

13   Gidley, D.W., Marko, K.A., and Rich, A., Phys. Rev. Lett.
     36, 395 (1976)

14   Ford, G.W., Sander, L.M., and Witten, T.A., Phys. Rev. Lett. 36, 1269 (1976)

15   Cherry, W., Ph.D. Thesis, Princeton Univ. 1958 (unpublished).

16   Canter, K.F., Mills, A.P., Jr., and Berko, S., Phys. Rev. Lett. 33, 7 (1974).

17   Gidley, D.W., Zitzewitz, P.W., Marko, K.A., and Rich, A, Phys. Rev. Lett. 37, 729 (1979).

18   Gidley, D.W., Rich, A., Zitzewitz, P.W., and Paul, D.A.L., Phys. Rev. Lett. 40, 737 (1978).

POSITRONIUM IN EXCITED STATES

E.W. Weber
Physikalisches Institut der Universität,
Philosophenweg 12, D-6900 Heidelberg, W. Germany

1. Introduction

Positronium is a bound pure leptonic particle-antiparticle system.
It can be described to high accuracy by the electromagnetic interaction
of the electron and positron and the interaction of both particles with
the photon fields. Effects due to the strong or weak interactions are
minimal. Therefore in principle positronium is ideal for a test of the
quantum electrodynamic theory of bound systems. Furthermore it allows
one to study real self-annihilation processes which have not yet been
investigated in any other system. Comprehensive review articles have
been published both on theory of and experiments with positronium by
Deutsch (1), De Bennedetti and Corben (2), and a mainly theoretical
paper by Stroscio (3). Articles covering more recent positronium
research are those by Mills, Berko, and Canter (4), Berko, Canter, and
Mills (5), Griffith and Heyland (6), and by Rich (7).

In practice, positronium (Ps) has several disadvantages both
theoretically and experimentally. On the theoretical side, the Ps atom
consists of two light equal mass particles which generally have to be
treated as a relativistic two body problem. Unlike atomic hydrogen it
cannot be reduced in a simple way to a one body problem with calculable
exact solutions by using the reduced mass concept. For Ps the
relativistic Bethe-Salpeter equation applies which, even in zero order,
has no exact solutions (3). A new approach circumventing this problem
was introduced by Lepage and Caswell (8), and Remiddi and Barbieri (9),
see Sec. 4. Furthermore, annihilation and two-photon terms contribute
in the same order of $R_\infty \alpha^3$ as the Lamb shift; higher order corrections
become exceedingly numerous and tedious to calculate.

Experimentally, more difficulties are encountered with Ps than with the stable hydrogen atom. Ps can be produced only in low abundance, even in its ground state, and decays with high annihilation rates. In addition, the annihilation gives rise to a background of gamma radiation. Because of its small mass and generally non-thermal energy, a huge linear Doppler effect and also a quadratic Doppler effect has to be expected in the case of optical transitions.

At the present time, the most accurate quantities measured for Ps are the $1^1S_0-1^3S_1$ ground state hyperfine structure (hfs) splitting (10,11) and the ortho-Ps ground state decay rate (12,13). In excited state positronium only one measurement, the fine structure (fs) separation $2^3S_1-2^3P_2$, has been performed to date (14). The present paper will deal mainly with experiments involving excited states of Ps with particular emphasis on the laser two-photon transition $1^3S_1-2^3S_1$ which seems feasible with recent developments in Ps production techniques (15,16) and in laser technology.

2.   Why experiments with excited states of positronium

The limiting factor in the determination of the ground state hfs splitting is the short 1^1S_0 lifetime of $1.25 \cdot 10^{-10}$s . To obtain an accuracy of 3 ppm the linewidth of the microwave resonance has to be split to 5 parts in 10^4. In comparison, resonances involving excited states of Ps (Table 1) have equal or smaller natural width to frequency ratios and thus may allow in principle considerably more accurate measurements. The fs transitions between n=2 Ps states also lie in the easily accessible microwave range. The qualities of excited Ps states for testing QED theory will be discussed in Sec. 4.

The most attractive and rewarding experiment would seem to be the measurement of the two-photon transition $1^3S_1-2^3S_1$ which has by far the smallest natural linewidth due to the annihilation lifetime of 1^3S_1 ($\tau(3\gamma)=1.4\cdot10^{-7}$s) for all transitions listed in Table 1. With the n>2 S states not being metastable*, transitions from 1^3S_1 to higher n^3S_1 states will not result in a smaller natural linewidth, although they have somewhat higher theoretical width to frequency ratios. The Ps Balmer series starting from the metastable

---

* The radiative lifetimes of Ps* states are about two times those of equivalent H* states; $\tau(Ps,n\ S)/10^{-7}$ s = 3.2(n=3),4.6(n=4),7.2(n=5).

Table 1. Theoretical width to frequency ratios of some transitions between n=1 and n=2 states in positronium.

Transition	Transition Frequency ν_o (GHz)	Dominant Decay	Decay Time τ(s)	Width Γ (MHz)	Width/ Frequency Γ/ν_o
$1^1S_0 - 1^3S_1$	203	$1^1S_0(2\gamma)$	$1.3\cdot10^{-10}$	1200	$6\cdot10^{-3}$
$2^3S_1 - 2^3P_2$	8.6	$2^3P(Ly-\alpha)$	$3.2\cdot10^{-9}$	50	$6\cdot10^{-3}$
$2^1S_0 - 2^3P_0$	18.5				$3\cdot10^{-3}$
$1^3S_1 - 2^3P_0$	$1.2\cdot10^6$	$2^3P(Ly-\alpha)$	$3.2\cdot10^{-9}$	50	$4\cdot10^{-8}$
$1^3S_1 - 2^3S_1$	$1.2\cdot10^6$	$1^3S_1(3\gamma)$	$1.4\cdot10^{-7}$	1.3	10^{-9}

n=2 $^3S_1$ state $(\tau(3\gamma)=1.1\cdot10^{-6}s)$ has the highest resolution because of the long 2^3S_1 annihilation time. This seems to be of academic interest to date, when considering the production rate ratio of order 10^{-3} for n=2 Ps* to ground state Ps from surfaces (17).

With the introduction of new Ps production methods and with present-day pulsed, tunable megawatt dye lasers the induction and detection of the $1^3S_1 - n^3S_1$ two-photon transitions seems feasible. The nonlinear two-photon laser spectroscopic technique (19) eliminates the first order-but not the second order- Doppler effect. The transverse Doppler effect prevents one from reaching the resolution theoretically expected from the natural width to frequency ratio for the Ps atom. This again arises from its small mass and consequently high velocity even at thermal energy. (Sec. 3.4). Nevertheless a determination of the $1^3S_1 - 2^3S_1$ splitting with an absolute accuracy of the order of 10 MHz, i.e., a $2\cdot10^{-8}$ relative error, seems possible in the near future.

3. Experiments with excited states of the positronium atom

3.1. Discovery of excited state positronium

Following the first observed generation of Ps in gases by Deutsch (20) in 1951 it took almost a quarter of a century before Ps n=2 state production was discovered. The search for Ps* had been pursued with different methods in many experiments. The only two approaches in which success has been reported will be described here.

In the first a group at Yale (21) used optical excitation of Ps atoms formed and stopped in 1 bar of Ar buffer gas from positrons emitted by a radioactive ^{22}Na source. Of the Ps formed, 75% existed in the longer lived 1^3S_1 state. A tin arc maintained in Ar carrier gas excited the Ps 1^3S_1 atoms to the 2^3P states making use of a fortuitous coincidence of a Sn line with Ps Ly-α at λ = 243 nm. A magnetic field of 70 mT was also applied to mix 2^3P with 2^1P states which decay to the 1^1S_0 ground state. The signal believed to have shown the excitation was an increase of the 1^1S_0 2γ annihilation decay rate by S(2γ)=0.149(24)%. The buffer gas stopping technique* has not been pursued any further for the Ps* research, because of the

---

* For a review of very fruitful e$^+$ and Ps ground state research in gas atmospheres see, e.g., refs. (22) and (6).

many disadvantages to be anticipated: high UV background from buffer
gas atoms or molecules excited by slowing down positrons; high
collisional quenching for the metastable n=2 Ps states; and strong
buffer gas density shift and broadening of the microwave and,
especially, the optical transitions.

<u>Slow positron beams and Ps formation on surfaces.</u>   The   second
approach to detect Ps* made use of the development of two new techniques: (1) the generation of slow positron beams of 10^4 to 10^7 e$^+$/s
intensities; (2) the conversion of slow positrons (several 10eV to keV
energies) to slow Ps atoms (with thermal to <4eV energies) at surfaces
of various materials.

Slow positrons (<5eV) are emitted into vacuum from various
surfaces with fractional intensities ranging from 10^{-7} to 10^{-3} compared
to those of the positrons incident from radioactive ^{22}Na or ^{58}Co
sources for which the mean kinetic e$^+$ energy is of the order of
0.5 MeV. The standard slow e$^+$ emitter has for a long time been MgO
fumed on gold foils lined up as "venetian blinds"; such an arrangement
has an efficiency up to $3\cdot10^{-5}$. More recent studies have shown that
clean single-crystals (Al, Cu, Si, Ge) and other surfaces in ultra-high
vacuum can reemit slow positrons with fractions up to 10^{-3} (17). The
slow e$^+$ are subsequently accelerated to some 10eV to keV energies and
then magnetically or electrostatically guided to the target area. The
construction and application of these slow e$^+$ beams is discussed in the
review articles (5-7).

Up to almost 100% of the positrons from a slow beam impinging on
various surface materials, e.g. MgO, SiO, Ti, Cu, Au, and W, can form
Ps atoms leaving the surface. A very efficient converter is a Cu (111)
single crystal surface covered with a 1/3 monolayer of S placed in an
ultra-high vacuum (15). This Cu crystal converts about 50% of the
incident e$^+$ into Ps with a thermal energy distribution having a mean
energy of $\bar{E}$ = 3.4(3)eV. Most of the remaining e$^+$ form Ps atoms which
are desorbed from the surface with $\bar{E}$ = 0.14(1)eV corresponding to the
T = 1060 K crystal temperature and with a non-Maxwellian velocity
distribution.

The Brandeis group used these two methods not only for the first
ob observation of Ps Lyman-α radiation (17) but also for a measurement of
the Ps(2^3S_1-2^3P_2) fs interval (14). Slow e$^+$ emitted from a MgO covered
gold foil converter are magnetically guided by a curved solenoid (7mT;

150 cm long) to a n-type Ge target where Ps atoms in the ground and excited states are formed (Fig.1). The fraction of n=2 state Ps atoms per incident positron emitted into the vacuum from the converter is on the order of 10^{-4}. The n=2 Ps atoms are detected via Ly-α photons in delayed coincidence with annihilation γ's from the 1^3S_1 decay.

3.2. Measurement of the positronium $2^3S_1-^3P_2$ fine structure interval

A level diagram of the Ps n=1 and n=2 states, with dominant decay lifetimes, is shown in Fig.2. Three of the 15 possible fs transitions are electric dipole transitions involving the long lived 2^3S_1 state. From those the fs interval $2^3S_1-2^3P_2$ at lowest (X-band) microwave frequency is chosen for the experiment (14). The measurement is based on the observation of an enhanced Lyman-α ($2^3P_2-1^3S_1$, λ = 243 nm) emission rate in delayed coincidence with subsequent 1^3S_1 annihilation γ's when the Ps* fs transition is induced by a microwave electric field at the appropriate frequency. A slow positron beam ($\approx$30eV) similar to that indicated in Fig.1 is magnetically guided into a cylindrical microwave cavity (Fig.3). The slow positrons collide with the Cu end face of the cavity and form n=2 Ps atoms leaving the surface at a fraction of 10^{-3} to 10^{-4} of the incident e^+ intensity. The subsequently emitted (2P-1S) Ly-α photons can leave the cavity through parallel wires replacing one of its sides and a quartz window; they are detected by a UV sensitive solar-blind photomultiplier. Two NaI(Te) detectors placed above and below the target chamber count the annihilation γ's. The time delay spectrum of the annihilation γ's is recorded following a Ly-α start signal. The increase of the long lifetime component of this spectrum with rf on is used for the detection of the desired transition sequence:

$$2^3S_1(\tau=1\mu s)\xrightarrow[+h\nu_{rf}]{} 2^3P_2(\tau=3ns)\xrightarrow[-h\nu_{Ly-\alpha}]{} 1^3S_1(\tau=140ns)+3\gamma \qquad (1)$$

The long component with $\tau\approx$120 ns of the delayed coincidence spectrum is shorter than the 1^3S_1 vacuum lifetime, which is attributed to wall collisions. Fuming MgO on the walls of the cavity indeed increases the long lifetime component significantly. This indicates that n=2 Ps atoms do not stick, nor are they strongly quenched on MgO surfaces, nor is the Ps* annihilation rate increased.

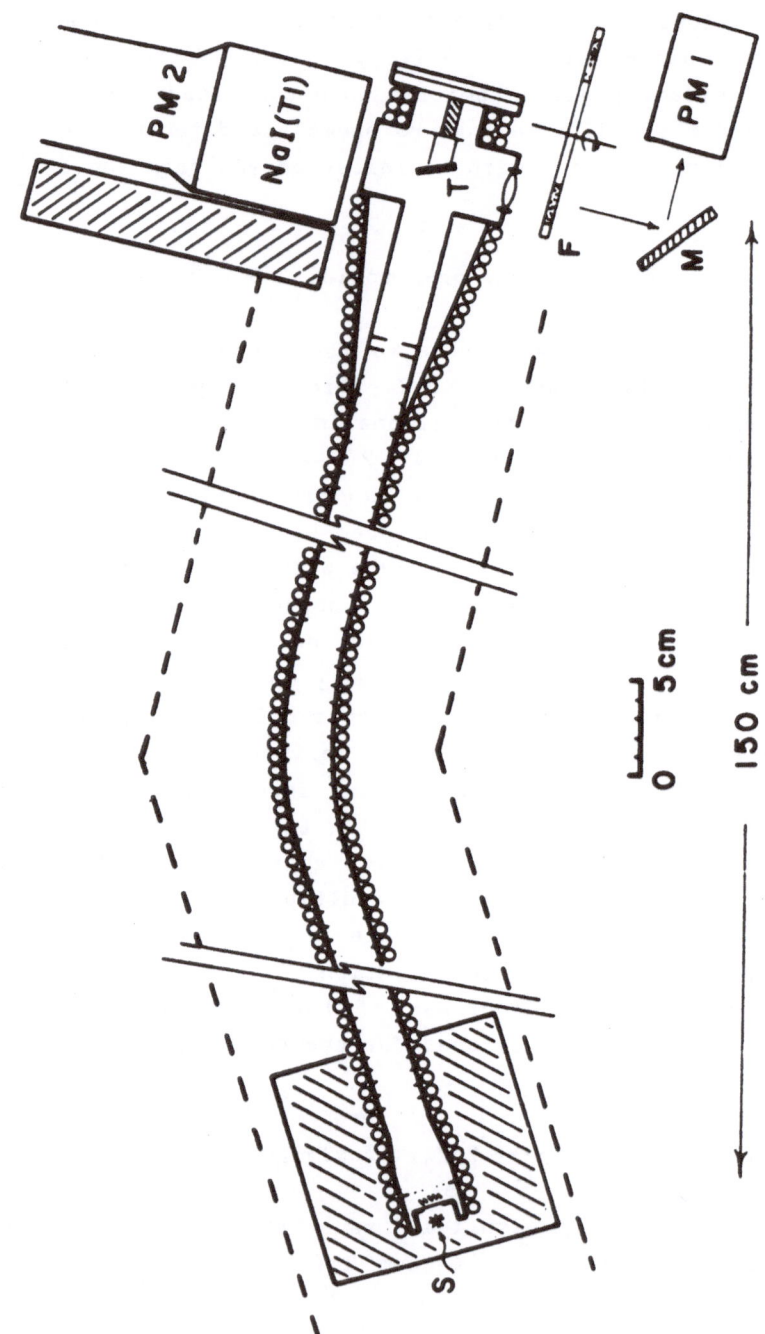

Fig.1. Slow positron beam apparatus for the detection of Ps Lyman-α radiation. S, [58]Co source; T, target; F, optical filter wheel; M, aluminized mirror; PMI, UV photon counter (from ref.(17)).

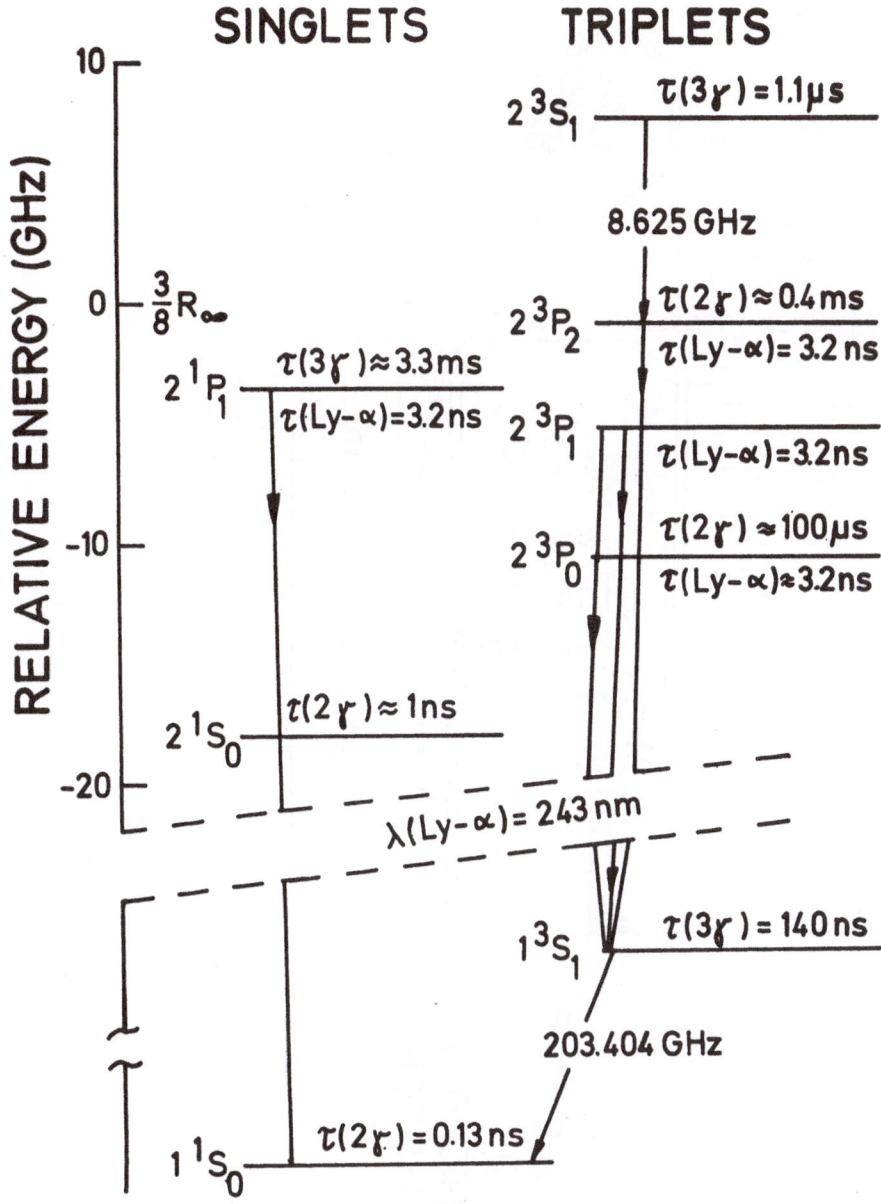

Fig.2. Level diagram of the positronium n=1 and n=2 states with approximate annihilation and radiative lifetimes (3,18).

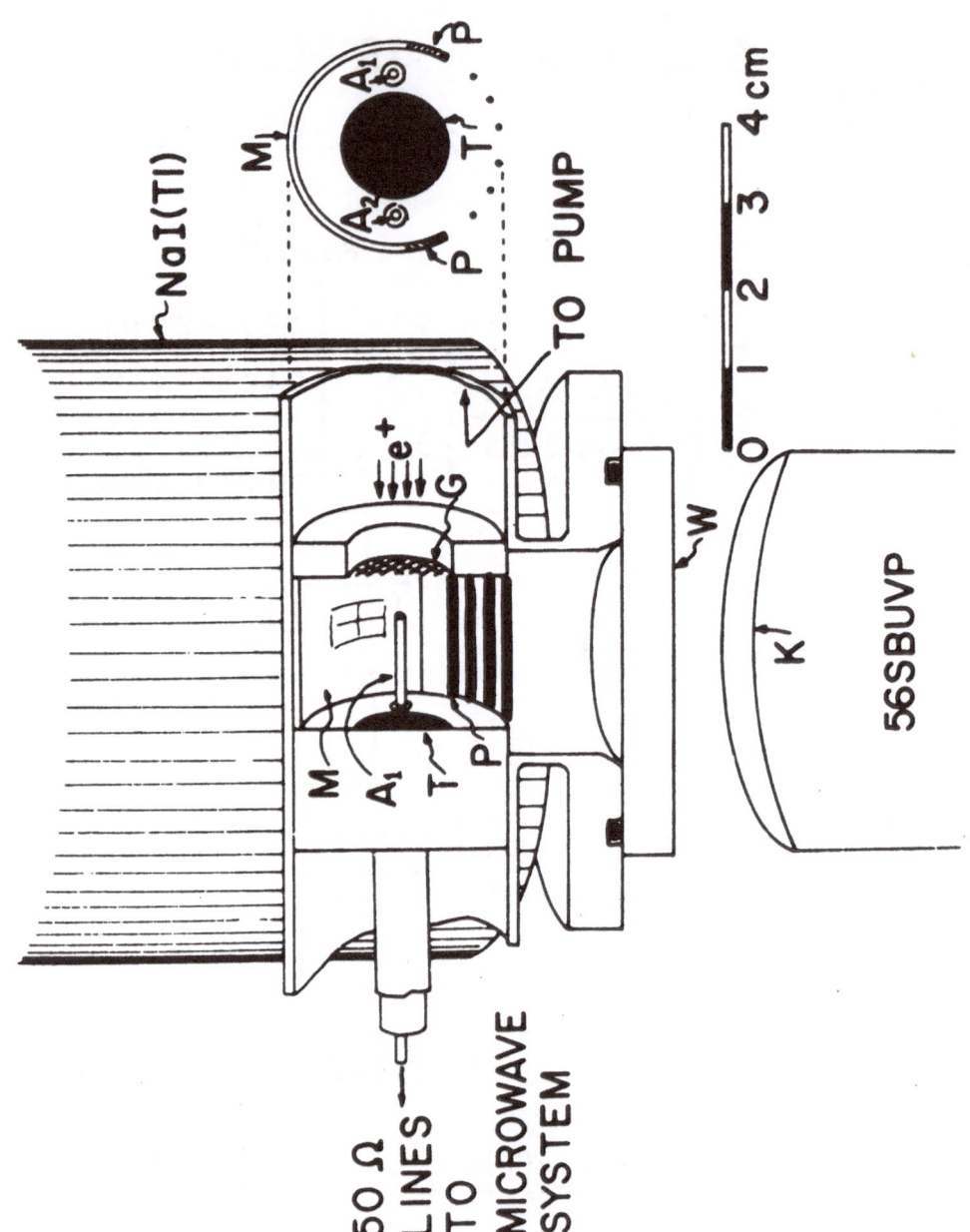

Fig.3. Experimental setup for the measurement of the positronium
($2^3S_1 - 2^3P_2$) fine structure interval with positron target chamber and
microwave cavity. G, 95%-transmission tungsten grid; T, copper target;
M, aluminized quartz mirror; W, quartz window; K, CsTe photocathode;
P, support posts; A_1, input antenna; A_2, output antenna; NaI(Tl),
annihilation detector (from ref.(14)).

The fs transition signal (Fig.4) is defined as

$$S(\nu_{rf}) = (N_{on}(\nu_{rf}) - N_{off})/N_{off}, \tag{2}$$

where N_{on} (N_{off}) are the counting rates in the long time delayed component of the (Ly-α - annihilation R) coincidence spectrum with the rf on (off) at frequency ν_{rf}. Also shown is the logarithmic first-difference signal

$$S'(\nu_{rf}) = [N(\nu_{rf}+\Delta) - N(\nu_{rf}-\Delta)]/[N(\nu_{rf}+\Delta) + N(\nu_{rf}-\Delta)] =$$
$$= [S(\nu_{rf}+\Delta) - S(\nu_{rf}-\Delta)]/[2+S(\nu_{rf}+\Delta) + S(\nu_{rf}-\Delta)] , \tag{3}$$

which is obtained in separate runs using a frequency modulation of the rf with amplitude $\Delta = \pm 30$MHz. An assumed Lorentzian line shape $S(\nu_{rf}) = 1/4 \cdot A\delta^2[(\nu_{rf}-\nu_o)^2 + 1/4 \cdot \delta^2]^{-1}$ is fitted (solid lines) to both curves S and S' yielding (14) A = 11.4(.6)%, δ = 102(12) MHz with χ^2/ν = 12.1/10 and $\nu_{expt}(2^3S_1 - 2^3P_2) = 8628.4(2.8)$MHz. The width δ is about twice the 50MHz natural width (Table 1) determined by the radiative decay of the 2^3P_2 state. This is in reasonable order of magnitude agreement with estimates of the detection geometry and efficiency, the rf power dependence and wall collision effects. The quoted uncertainty is the pure statistical error.

For a comparison with theory the experimental value has to be corrected for two effects*. The motional Stark shift (23) due to a residual guiding magnetic field of about 5 mT results in an estimated +3(2)MHz correction for Ps atom energies ranging between 0 and 1 eV. A possible variation of the microwave electric field strength with frequency may add a few MHz to the error. A reasonable value for the corrected result seems to be

$$\nu_{expt}^{corr}(2^3S_1 - 2^3P_2) = 8631(5) \text{ MHz .}$$

This must be compared with the theoretical value (calculated by Fulton and Martin (24) in 1954) which includes all radiative corrections to order $R_\infty\alpha^3$(corresponding to order $mc^2\alpha^5$),

$$\nu_{theor.}(2^3S_1 - 2^3P_2) = 8625.14 \text{ MHz.}$$

---

* This deviates from the procedure chosen in ref.(14) where the corrections are applied on the theoretical value.

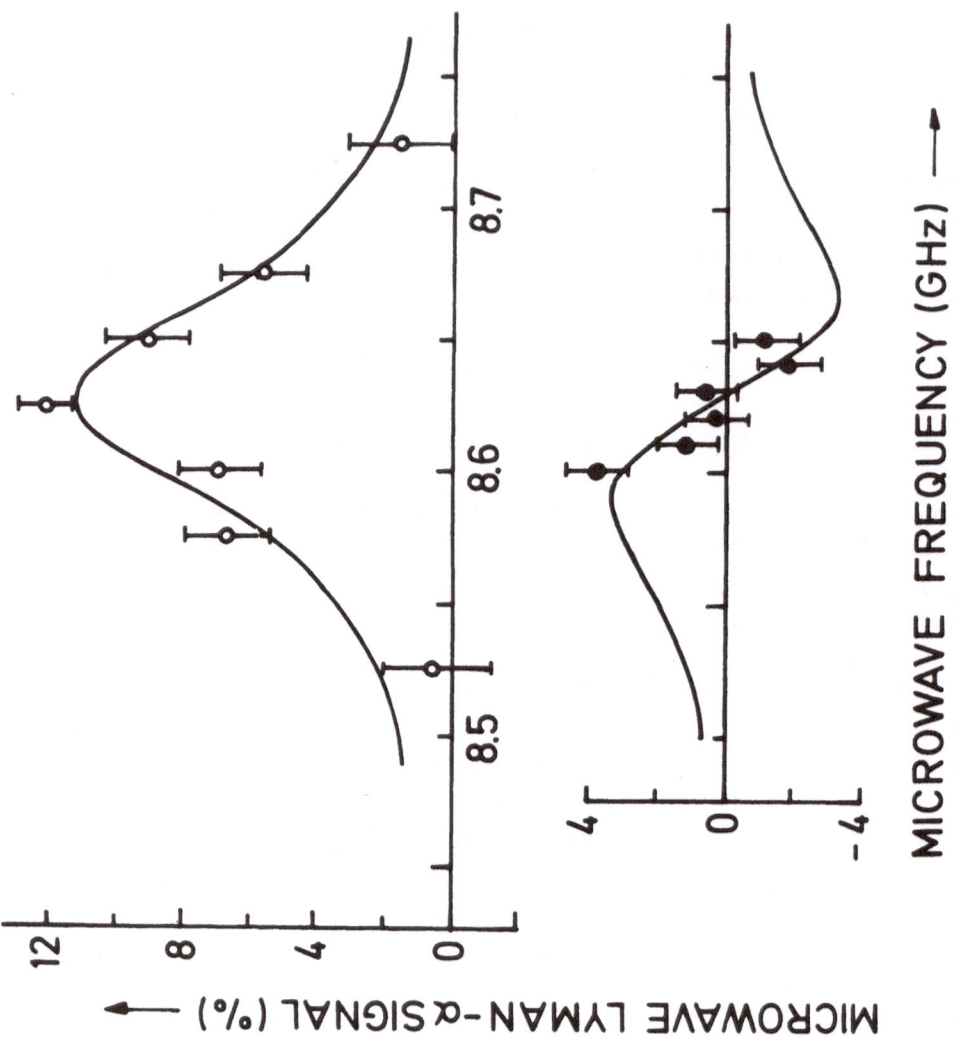

Fig.4. Positronium $(2^3S_1 - 2^3P_2)$ fine structure transition signal S(open circles) observed as a delayed (Lyman-α - annihilation γ) coincidence signal as a function of microwave frequency. Also shown is a logarithmic first-difference signal S'(solid circles) obtained by frequency modulation of the rf (from ref.(14)).

The experimental and theoretical values can be considered to be in fair agreement when taking into account contributions of the order of MHz to the theoretical value from higher order terms and the uncertainty in the estimates of the systematic corrections of the experimental value. The measurement already confirms the $R_\infty \alpha^3$ radiative correction terms (Lamb shift, virtual annihilation, and two photon interaction, see Fig.7) of 231 MHz to a few percent accuracy. A new experiment at Brandeis can be expected to achieve a 1 MHz accuracy. Comparison with theory will then require the evaluation of the higher order, $R_\infty \alpha^4 \ln \alpha^{-1}$ and $R_\infty \alpha^4$ terms.

In order to approach the few ppm precision achieved for the positronium or muonium ground state hfs measurements, both the systematic and statistical uncertainties had to be reduced by two orders of magnitude. With half the present experimental error originating from counting statistics for running times of a day, the e^+ beam intensity and the fraction of 2^3S_1 Ps atoms produced directly (or excited from the 1^3S_1 state, see Sec.3.4) has to be significantly improved. The natural width of 50 MHz can practically be approached by replacing the magnetic guiding field by an electrostatic system and by using the recently developed thermal Ps production techniques (15,16) to reduce motional Stark shift and Doppler broadening. A factor of two in the linewidth to frequency ratio (Table 1) can readily be gained by switching to the $2^3S_1 - 2^3P_0$ fs interval at a frequency of 18,496 MHz. A possible scheme for efficient $1^3S_1 - 2^3S_1$ excitation and detection of n=2 fs microwave transitions is described at the end of Sec.3.4.

3.3. Feasibility of positronium Lyman-α spectroscopy

The classical optical spectroscopy is limited in resolution by the Doppler linewidth. For the light positronium atom the Doppler width given by $\Delta \nu_D = 7 \cdot 10^{-7} \nu_0 \cdot \sqrt{T/M}$ is of the order of 700 GHz for Ly-α and T=800 K. The linewidth to frequency ratio is $6 \cdot 10^{-4}$, and neither the n=2 fs nor the n=1 hfs intervals can be resolved. For resolution of the optical transitions in the Ps atom one has to resort to Doppler-free, nonlinear laser spectroscopic techniques, therefore.

Discussion of the feasibility of positronium Lyman-α saturation experiments. Saturated absorption resonances are limited in resolution only by the natural lifetimes of the states involved. Here again the ortho-Ps system has an advantage over para-Ps because of the longer annihilation lifetime of the 1^3S_1 ground state (Fig.2). The natural line-width of the 1^3S_1-2^3P transitions of 50 MHz is determined by the radiative lifetime of the ^3P excited states resulting in a width to frequency ratio of $4 \cdot 10^{-8}$ (Table 1). Because of the Ly-α photon absorption recoil the Ps saturation resonances will be split into two lines separated by about 6 GHz which is much greater than the natural width.

Only those Ps atoms will contribute to the narrow saturation resonance which have nearly zero velocity components in the direction of the counterpropagating laser beams. A Doppler shift $\Delta\nu = \nu_o(v/c)\sin(\pi/2-\vartheta)$ of the order of the natural halfwidth of 25 MHz is seen by a Ps atom of mean velocity $v(T=800K)=10^7$cm/s moving at an angle $(\pi/2-\vartheta)=6 \cdot 10^{-5}$ rad off from perpendicular to the laser beams (Fig.5a). A fraction of only $3 \cdot 10^{-9}(3 \cdot 10^{-5})$ of the Ps atoms will contribute to a 50 MHz (5 GHz) wide Lamb-dip signal when Knudsen's law, $dn \propto \cos\theta \, d\theta$, (25) is assumed to approximately describe the angle dependence of the Ps atom desorption from a $(e^+\text{-Ps})$ converter surface. There is, furthermore, no obviously suitable, highly sensitive method available to detect the Ps Ly-α saturated absorption resonances. The Lamb-dip detection method (26) via reduction of the 3P state 2γ annihilation suffers from large background and solid angle problems. In conclusion, the observation of narrow Ps Ly-α Lamb-dip signals seems problematic* with present-day Ps fluxes of $\leq 10^7$/s having little or no angular resolution, and with the difficulties associated with the generation of long-lasting, intense, pulsed, tunable UV laser sources at $\lambda=243$ nm.

3.4. Possible two-photon laser spectroscopy of the positronium 1^3S_1-2^3S_1 transition

In the case of positronium the nonlinear two-photon spectroscopy has three important advantages compared to saturated absorption spectroscopy**: (i) The required laser radiation lies in the blue wave-

---

* For a more optimistic estimation, see ref. (26).

** A similar experiment with respect to the laser spectroscopy method has been successfully performed for the 1S-2S two-photon transition in atomic hydrogen by Hänsch and coworkers (27). The transition is detected by Ly-α ($\lambda=121.5$ nm) radiation following a (2S-2P) quenching collision in the hydrogen gas. In the latest version of this pioneering experiment the 1S-2S interval is measured with an accuracy of $1.2 \cdot 10^{-8}$ relative to the H_β ($4S_{1/2}$-$4P_{1/2}$, $S_{1/2}$ crossover transition induced by saturated polarization spectroscopy (28).

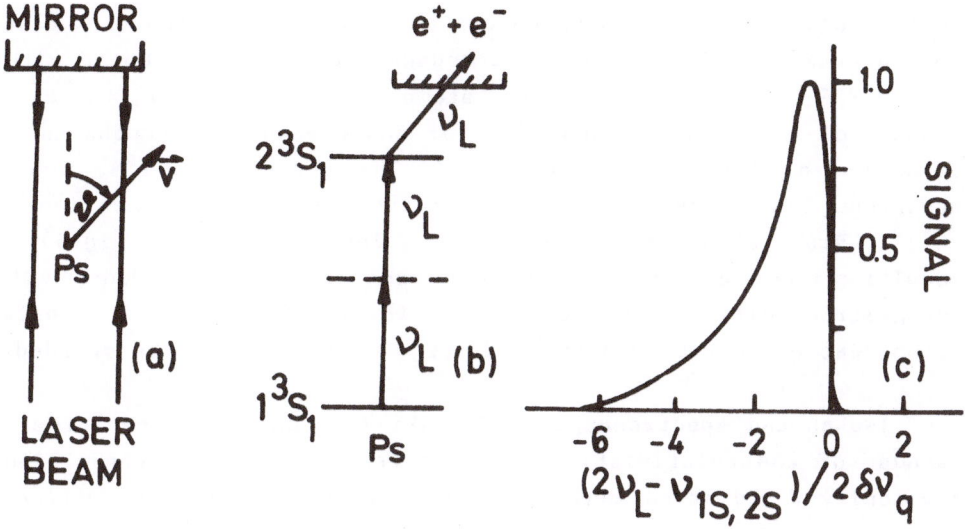

Fig.5. Laser two-photon absorption scheme for positronium, with (a) counterpropagating λ = 486 nm laser beams, (b) two-photon excitation $1^3S_1 \rightarrow 2^3S_1$ and subsequent photoionization Ps $2^3S_1 \rightarrow e^+ + e^-$, and (c) two-photon resonance signal, which is shifted and asymmetrically broadened due to the quadratic Doppler effect.

length region at $\lambda=486$ nm which can be readily generated with a tunable pulsed Coumarin 102 dye laser at megawatt powers when pumped with a N_2 or Nd-YAG laser. (ii) All Ps atoms will contribute to a single two-photon transition independent of their velocity components in the direction of the laser beams. (iii) The narrow $1^3S_1-2^3S_1$ two-photon resonance can be detected very efficiently by photoionization of the Ps 2^3S_1 state with the same $\lambda=486$ nm laser photons (Fig.5). The resulting free e^+ and e^- are then recorded in coincidence with two channeltron multipliers (Fig.6). Thus the annihilation γ detection techniques which are subject to a large background can be avoided.

Two-photon spectroscopy is Doppler-free only in first order (19). Because of the relativistic Doppler effect the Ps atom travelling with a velocity v and at an angle ϑ relative to the laser beams will see two frequencies ν_+ and ν_- from the counterpropagating beams of frequency ν_L (Fig.5),

$$\nu_\pm = \nu_L \frac{1\pm\frac{v}{c}|\cos\vartheta|}{(1-(\frac{v}{c})^2)^{1/2}} \approx \nu_L(1\pm\frac{v}{c}|\cos\vartheta|+\frac{1}{2}(\frac{v}{c})^2\pm\ldots).\tag{4}$$

When the Ps atom absorbs one photon from each of the laser beams the sum frequency is given in second order by

$$\nu_o = \nu_+ + \nu_- \approx 2\nu_L(1+\frac{1}{2}(\frac{v}{c})^2)\tag{5}$$

The quadratic Doppler effect thus causes a red shift $\delta\nu_q = \nu_L(\frac{v}{c})^2\approx70$MHz and an asymmetric broadening ≈ 250 MHz (26) for Ps atoms with a (Maxwellian) velocity distribution and a (mean) velocity of 10^7 cm/s. With a linewidth to frequency (ν_L) ratio of $4\cdot10^{-7}$ and an assumed 5% accuracy for the determination of the (asymmetric) line center, a relative error of $2\cdot10^{-8}$ seems to be attainable, even without a complete knowledge of the exact Ps velocity distribution (15). The necessary frequency stability and narrow bandwidth can be obtained by using a single-frequency cw dye laser as input of a pulsed dye laser amplifier system as described in ref. (28). At the 10^{-8} level of accuracy, the wavelength can either be measured with a digital wavemeter (29), a calibrated interferometer (30), or relative to the wavelength of a Balmer-β fs saturation transition in atom hydrogen (compare (27, 28)).

The transition rate for two-photon excitation is given by (31)

$$R_{1S,2S} = \frac{1}{2h^2\Gamma} \left| \sum_n \frac{V_{2S,n} \, V_{n,1S}}{h(\nu_L - \nu_{n,1S})} \right|^2 \cdot \frac{\Gamma_{2S}}{\Gamma_2} \cdot L(2\nu_L - \nu_{1S,2S}) \tag{6}$$

where $V_{n,1S}$ and $V_{2S,n}$ are the electric dipole matrix elements from intermediate states n to the 1S and 2S states, $\nu_{n,1S}$ are the frequencies of the virtual intermediate transitions; $\nu_{1S,2S}$, Γ, and Γ_2 are the frequency, total width, and homogeneous width of the two-photon transition, respectively; Γ_{2S} is the width of the upper 2S level. The line shape function obtained through the Ps velocity distribution is denoted by L. The transition rate $R_{1S,2S}$ can be calculated to good approximation by analogy to hydrogen (32), taking into account the facts that for Ps the energies are half, and the radial extensions and lifetimes are twice those for hydrogen.

With the assumption of a Maxwellian velocity distribution for the Ps atoms and with $\Gamma_2/\delta\nu_q \ll 1$ (which is the case) the line shape function becomes (26),

$$L(2\nu_L - \nu_{1S,2S}) \approx \frac{\Gamma_2}{4\sqrt{2\pi}\delta\nu_q} \exp\left(\frac{2\nu_L - \nu_{1S,2S}}{2\delta\nu_q} \right) \times \tag{7}$$

$$\times \left\{ \left[\left(\frac{2\nu_L - \nu_{1S,2S}}{2\delta\nu_q} \right)^2 + \left(\frac{\Gamma_2/2}{2\delta\nu_q} \right)^2 \right]^{1/2} - \frac{2\nu_L - \nu_{1S,2S}}{2\delta\nu_q} \right\}^{1/2}$$

with the maximum value of $L = \Gamma_2/4\sqrt{2\pi e} \cdot \delta\nu_q \approx 10^{-3}$ obtained for $2\nu_L - \nu_{1S,2S} = -\delta\nu_q$.

By using σ^+ or σ^- polarized laser light only those photons absorbed simultaneously one from each of the counter-propagating beams can contribute to the 1S-2S transition. This is assumed for the derivation of Eq.(7). For a laser pulse of duration $\Delta\tau_L$, the two-photon transition probability is obtained as

$$w_{1S,2S} \approx 4 \cdot 10^{-3} \, I^2 \cdot \Delta\tau_L \cdot L(2\nu_L - 2\nu_{1S,2S}) \tag{8}$$

with I being the mean laser intensity during a pulse in W/cm^2. For a single laser pulse of $\Delta\tau_L = 10$ ns duration, a power of 1 MW, and a beam diameter of 10 mm (expanded from an original 1 mm), the intensity of $1.3 \, MW/cm^2$ is sufficient to obtain $w_{1S,2S} = 1$.

The ionization probability w_{ion} of the Ps 2^3S_1 state in a standing wave laser field (factor 2 in the intensity!) is given by

$$w_{ion} = 2 \cdot \frac{I}{h\nu_L} \cdot \Delta\tau_L \cdot \sigma_{2S} \qquad (9)$$

The absorption cross section σ_{2S} for photoionization from the 2S state can be calculated from (33),

$$\sigma_{2S} = \frac{8\pi^3 e^2 \nu}{c} |V_{W,2S}|^2, \qquad (10)$$

where $V_{W,2S} = \int u_W^* \sum x_i u_{2S} \, d^3x$ is the integral over the wavefunctions of the continuum state, u_W, and the 2S state, u_{2S}, describing the dipole matrix elements; x_i are the directions of the photon polarizations. The kinetic energy W of the free electron plus positron is related to the energy $h\nu_L$ of the ionizing photon by $W = (h\nu_L - hR_\infty/8)$ with $hR_\infty/8$ being the Ps 2S ionization energy, $\nu_L = 3R_\infty/16$, and R_∞ the Rydberg constant in frequency units (34).

The dipole matrix elements can be solved in closed form analogous to those for hydrogen (35)

$$|V_{W,2S}|^2 = \frac{2^{17} \exp(-(4/r)\arctan 2r)}{1 - \exp(-2\pi/r)} \cdot \frac{(1+r^2)}{(1+4r^2)^6} \cdot \frac{a_o^3}{e^2} \qquad (11)$$

The probability of ionization from the Ps 2S state for laser pulses of wavelength $\lambda = 486$ nm is then obtained as

$$w_{ion} \approx 90 \; I \cdot \Delta\tau_L \qquad (12)$$

which results in $w_{ion} \approx 1$ for a single laser pulse having the above given properties.

A possible scheme for a positronium two-photon experiment is shown in Fig.6. A pulsed dye laser system is used to induce the transitions (Fig.5)

$$Ps \; 1^3S_1 \xrightarrow{+2h\nu_L} Ps \; 2^3S_1 \xrightarrow{+h\nu_L} e^+ + e^- + W \qquad (13)$$

with the free electron and positron counted in coincidence by the two channeltrons. For the triggering of the pump and/or dye lasers two schemes seem to be suitable. When positrons from a slow beam (Sec.3.1.) hit the cone of a channeltron, a fraction of 15% pick up an e^- to form Ps and simultaneously produce secondary electrons, which are also ejected from the surface (36). These can be used to give a channeltron trigger signal, (a) in Fig.6, for the Laser pulse, with a technically feasible 100 ns delay ±30 ns jitter. The second possibility is to use the method developed by Mills (37) to extract bunches of positrons out of a slow e^- beam apparatus. The delay time between the prestart pulse

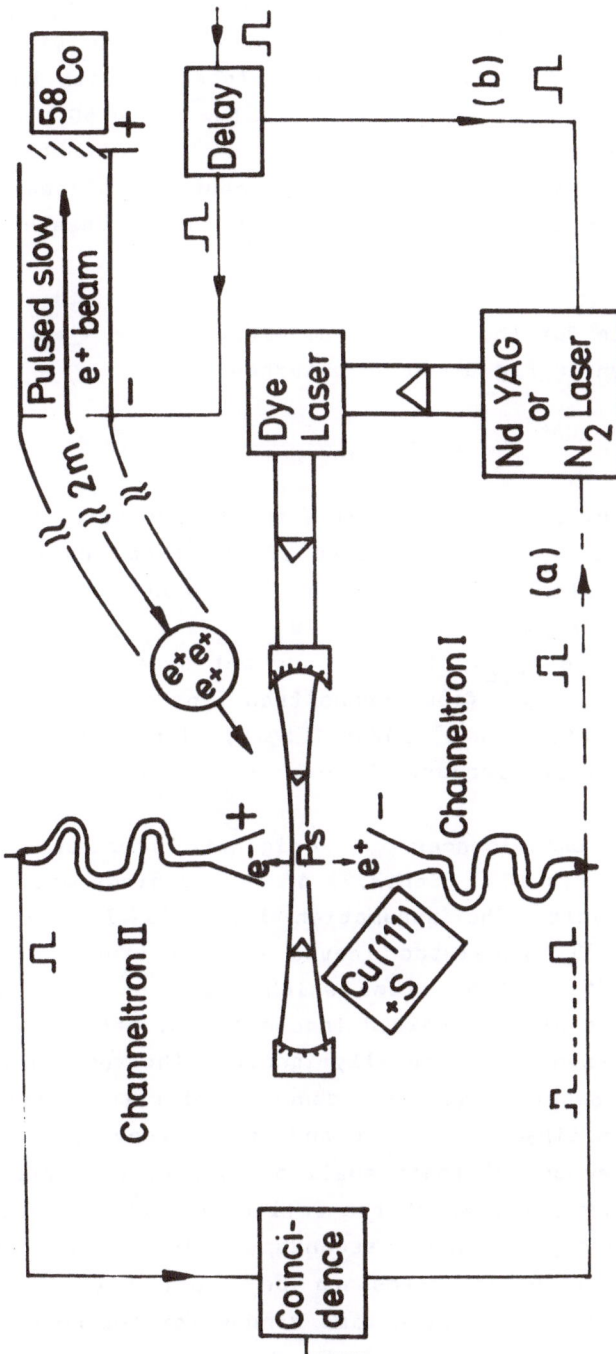

Fig.6. Possible positronium two-photon experiment with a pulsed, tunable megawatt dye laser. For triggering the laser either the channeltron I pulse (a) following Ps atom formation or the delayed slow positron bunch pulse (b) can be used.

of the laser, (b) in Fig.6, and the e^+ bunching pulse could be optimized for laser start jitter and for Ps time of flight from the converter surface into the laser beam. Together with the bunch method, the Cu(111) crystal converter (15) for slow e^+ to 50% thermal Ps (Fig.6 and Sec.3.1) can be favourably applied. The bunched production of thermal Ps atoms is very attractive because it allows one to trigger the laser before the Ps burst and thus to compensate for trigger to laser pulse delays.

An estimate for the (e^+,e^-) coincidence rate $R_{e^+e^-}$ detectable with the scheme shown in Fig.6 can be obtained from

$$R_{e^+e^-} = \frac{1}{T} \frac{\Omega}{2\pi} e^{-\eta \Delta t} \cdot w_{1S,2S} \cdot w_{ion} \tag{12}$$

where T^{-1} = 20 Hz is the laser pulse repetition rate, $\Omega/2\pi \approx 0.1$, the solid angle seen by the laser beam of the (one) Ps atom emitted from the converter surface, $\eta = \tau^{-1}(1^3S_1 \to 3\gamma)$, and Δt = (100±30)ns is a laser pulse delay and/or jitter time. For $w_{1S,2S} \approx w_{ion} \approx 1$ one could expect a coincidence rate $R_{e^+e^-}$ of about one per second. This estimate holds for one Ps atom trigger from channeltron I or one Ps atom formed on the Cu(111) surface after an e^+ burst (Fig.6). The reason for using only a few e^+ in a bunch is discussed below.

Background coincidences can originate from several channels. A fraction of 10^{-3} to 10^{-4} (Sec.3.2) of the Ps atoms are formed in all 6 excited n=2 levels. The assumption that all 16 partly degenerated states are equally populated leaves only 3 states or a fraction of $2 \cdot 10^{-4}$ to $2 \cdot 10^{-5}$ Ps 2^3S_1 atoms which live long enough to reach the laser beam and to become ionized independently off the laser frequency. This small fraction can be totally ignored. The annihilation γ's will not generate a substantial coincidence count rate because the channeltrons see only a tiny solid angle and are, moreover, practically transparent to γ's because of their small mass. A more severe problem arises from the slow positrons which are reflected from the converter surface and simultaneously produce secondary electrons. For the Cu(111) converter this fraction seems to be small. Its influence can be minimized by setting a delayed time window for the coincidence equal to the time of flight of thermal Ps to and through the laser beam. Furthermore the channeltrons will produce a coincidence count independent of the number of e^+ and e^- stemming from one or several Ps atoms simultaneously ionized. Therefore the minimum number of e^+ per bunch can and should be used which will ensure one or only a few Ps atoms formed per burst. With an assumed 50% coincidence background count rate it should be possible to record an asymmetric two-photon resonance curve of 250 MHz width with a statistical signal to noise ratio of 8 to 1 within an hour. This is sufficient for a determination of the resonance maximum with an accuracy of 12 MHz or 5%. In conclusion, the described two-photon Ps experiment seems feasible with present-day lasers and Ps production techniques.

There are several possible improvements which have not been accounted for in making the estimates given above. The laser intensity and pulse length seen by the Ps atoms can be increased by using an external cavity (Fig.6) for storing the light. The Ps ground state atoms formed can be stored by reflecting them from MgO covered walls of a box surrounding the laser beams. Furthermore, a new idea for trapping excited neutral atoms, including Ps*, having positive Stark energies has been proposed by Wing (38). However, the applicability and usefulness of such an electrostatic trap (requiring high electric fields (up to 100 KV/cm) and low energies of the particles to be trapped (<10μeV)) for Ps* research remains to be seen.

The same excitation and detection scheme can be applied to observe microwave induced 2^3S_1-2P fs transitions. With $w_{1S,2S}$ and w_{ion} close to 1 for each laser pulse, the transitions will show up as a reduction of the(e^+,e^-)coincidences counting rate. With this techniques a determination of the Ps n=2 fs intervals seem possible with an accuracy higher by at least one order of magnitude compared with the experiment described in Sec.3.2.

4.   Status of the theory for positronium energy levels

Positronium is particularly suited for a test of the relativistic, bound, two-body problem in QED theory. The pure leptonic particle-antiparticle system will exhibit minimal effects due to strong or weak interactions. On the other hand, self-annihilation processes play an important role.

In general, the relativistic Bethe-Salpeter equation has been used in the calculations of the positronium energy levels. This equation has the disadvantage that it has no exact solutions, even in zero order, and that the zero'th order wave functions are infinite at the origin. They can be made finite only after a complicated renormalization

procedure including all orders of the fine structure constant α. Therefore higher order radiative corrections to the Bethe-Salpeter equation
are difficult to calculate with perturbation theory and to analyse and
compute numerically. A comprehensive review article covering the
theoretical developments until 1975 has been written by Stroscio (3).

4.1. Recent advances

A new two-body formalism for positronium and muonium has more
recently been introduced by Lepage and Caswell (8) , and Remiddi and
Barbieri (9). The Bethe-Salpeter equation is reduced to an equivalent
one-body Dirac equation or, in the non-relativistic case, to an
equivalent Schrödinger equation with reduced mass, by placing one of
the particles effectively on the mass shell. For these equations exact
analytical solutions exist in zero order, which considerably simplifies
the analysis and evaluation of higher order terms through perturbation
theory. In particular, the zero order wave functions remain finite at
the origin and reduce to the atomic hydrogen functions in the non-relativistic case. For a more in depth discussion of this new formalism,
the reader may refer to the review of Lepage (39) and to the article by
E. Borie (40) in this volume. In the case of the corrections to the
ground state fs interval all terms of order* $R_\infty\alpha^2$ and $R_\infty\alpha^3$ have been
calculated. By means of the new approach most of the $R_\infty\alpha^4 \ln\alpha^{-1}$ and $R_\infty\alpha^4$
terms have been evaluated, as well. The corrections to the excited
states intervals were calculated to order $R_\infty\alpha^2$ by Ferrell (41) and to
order $R_\infty\alpha^3$ by Fulton and Martin (24). For the absolute shifts of ground
and n=2 excited states, which are important for the optical laser
experiments, not all contributions of order $R_\infty\alpha^3$ have been evaluated.

4.2. Comparison between atomic hydrogen and positronium

To give an impression of the differences between positronium and
the hydrogen atom, the Feynman graphs contributing in orders R_∞, $R_\infty\alpha^2$,
and $R_\infty\alpha^3$ to the calculation of energy levels are shown in Fig.7, with
the corresponding numerical values listed in Table 2. The most
important features are the following: The J degeneracy is lifted in

---

* The terms $R_\infty\alpha^n$ correspond to $mc^2\alpha^{n+2}$ in another nomenclature, with
the Rydberg constant $R_\infty = \alpha^2/2\lambda_c$ and the Compton wavelength of the
electron $\lambda_c = h/mc$.

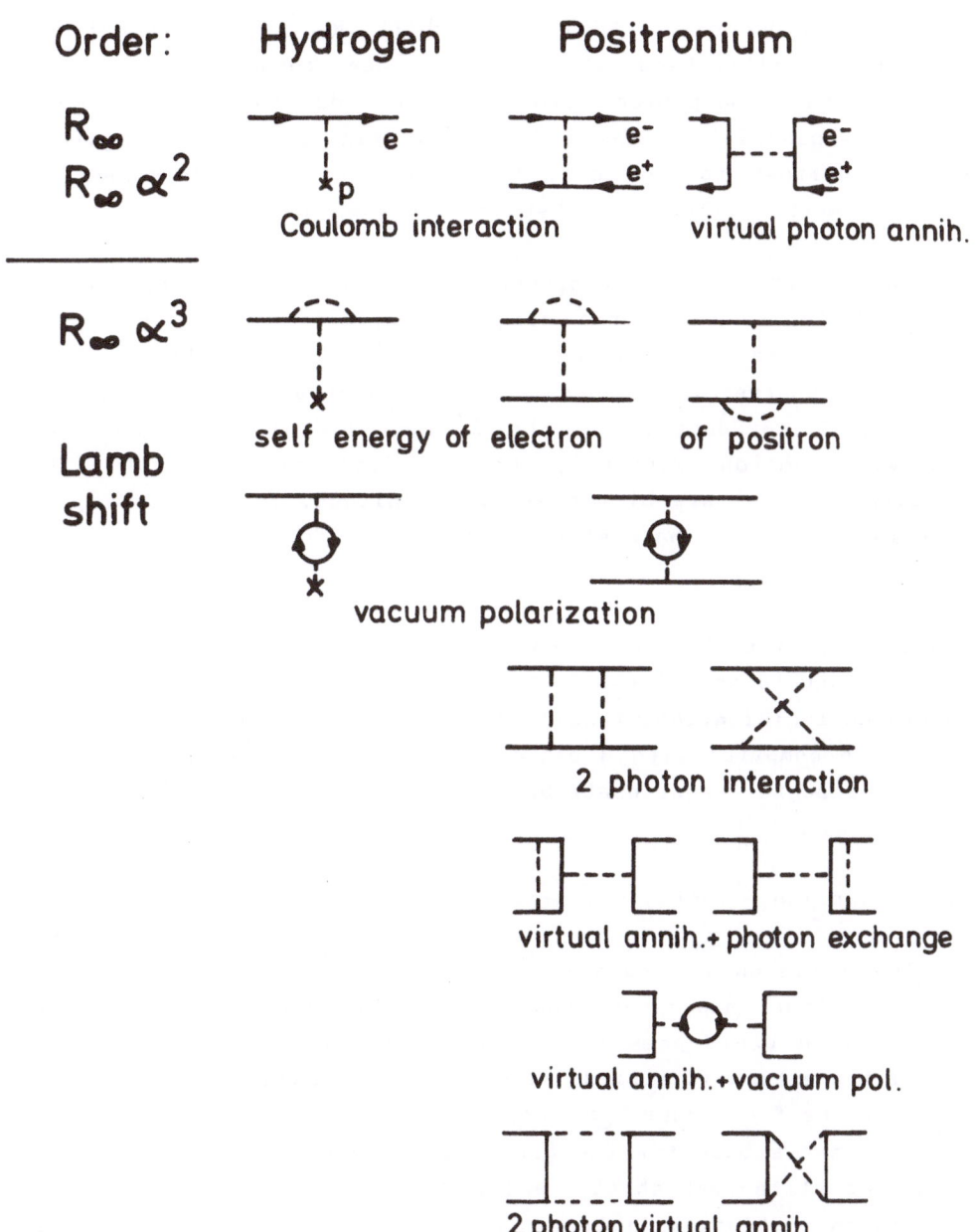

Fig.7. Comparison of Feynman graphs contributing to the hydrogen and positronium atom energies in the order of $R_\infty\alpha$, $R_\infty\alpha^2$, and $R_\infty\alpha^3$.

hydrogen only due to the famous Lamb shift $R_\infty \alpha^3$ radiative corrections. For Ps, on the other hand, it has already been removed in the order $R_\infty \alpha^2$ due to virtual one-photon annihilation and due to fs (spin-orbit) and hfs (spin-spin) interactions. The latter two are of same magnitude for Ps, in contrast to hydrogen, for which the magnetic moment of the proton is much smaller than that of the electron.

In the order $R_\infty \alpha^3$ ten graphs contribute for Ps compared to only two for H. The Ps $R_\infty \alpha^3$ terms originate primarily from contact inter-actions. Therefore the largest shifts from 15 to 294 MHz are observed for the 2S states. In the case of the 2P states the contributions from recoil (pair production in the Coulomb field) retardation effects (transverse photon exchange), and the classical Lamb shift (the same for all P states) are of comparable magnitude. In this order, virtual annihilation induces no P state shift.

Higher order correction terms become exceedingly numerous and complex for the Ps atom and therefore tedious and difficult to enumerate and to calculate. On the other hand, new and more information concerning annihilation, recoil and retardation effects can be obtained from Ps as compared with H since these effects show up in lower order and give rise to larger contributions.

4.3. Discussion - theory vs. experiment

The envisioned accuracies of $2 \cdot 10^{-8}$(12MHz) for the $(1^3S_1 - 2^3S_1)$ two-photon laser experiment and of $3 \cdot 10^{-5}$(0.5 MHz) for the $(2^3S_1 - 2^3P_0)$ fs transition would pose a substantial challenge for theory. For the optical transition it will be necessary to calculate all absolute shift contributions from order $R_\infty \alpha^3$ terms and most likely also from order $R_\infty \alpha^4 \ln \alpha^{-1}$ terms both for the ground and excited states of the Ps atom. The ground state Lamb shift, recoil, retardation, and two-photon anni-hilation corrections will be tested at the 10^{-2} level of accuracy. For the latter three radiative corrections this is considerably better than is presently tested for atomic hydrogen. Furthermore the mass ratio of the electron and positron can be obtained with $2 \cdot 10^{-8}$ accuracy yielding a high precision test of the CPT conservation law for the electromag-netic interaction. The rest mass of the positron will then be absolute-ly known with the same error $\Delta m/m = 5 \cdot 10^{-6}$ (42) as for the electron.

Table 2: Corrections to the positronium n=2 state non-relativistic energy $(3/8)hR_\infty$ from terms of order $R_\infty\alpha^2$[41] and $R_\infty\alpha^5$[24] in units of MHz.

State	$R_\infty\alpha^2$ shift fs + hfs interaction virtual photon annih.	self energy, vacuum pol.	$R_\infty\alpha^5$ shift pair product., photon exch.	virtual annih.	Sum
2^1S_0	-18,135	112	260	- 15	357
2^3S_1	7,413	61	294	-124	232
2^1P_1	- 3,536	-3	0	0	- 3
2^3P_2	981	-3	4	0	1
2^3P_1	- 5,360	-3	- 2	0	- 5
2^3P_0	-10,835	-3	-13	0	-16

In the case of the different (2S - 2P) fs transitions all $R_\infty \alpha^4 \ln \alpha^{-1}$ and also $R_\infty \alpha^4$ terms have to be evaluated. This will pose a stringent test of the large 2S state radiative corrections, and the identical Lamb shift but different recoil and retardation terms for the $2\,P_J$ states. With the excitation of the classical Lamb shift, the radiative corrections will be tested to higher accuracy than in any other bound system or scattering experiment to date. In principle, high-precision experiments with low energy transitions in the Ps atom will test the QED theory of radiative corrections for small distances between the particles involved and/or for short times of interaction. They are, therefore, partly equivalent to difficult and often low precision experiments at high energies where many other interaction channels are open, as well. The measurement of optical transitions in Ps will certainly motivate higher order calculations not only of the small energy differences between states of same principal quantum number, but also the evaluation of absolute shift corrections. The mastery of the theoretical treatment of this simple system for which the interaction is understood will supply some of the implements for studying and learning more about the properties of and forces operating in heavier particle-antiparticle systems.

Acknowledgements

The author would like to thank P.W. Zitzewitz for helpful discussions and R.M. Herman for valuable comments and carefully reading the manuscript.

References:

1. M.Deutsch, Prog. Nucl. Phys.$\underline{3}$, 131(1953)

2. S.De Bennedetti and H.C. Corben, Ann.Rev. Nucl.Sci.$\underline{4}$, 191(1954)

3. M.A. Stroscio, Phys.Rept. $\underline{22}$C, 215(1975)

4. A.P. Mills, Jr., S. Berko, and K.F. Canter, Atomic Physics 5, 103(1977)

5. S. Berko, K.F. Canter, and A.P. Mills, Jr., Prog.Atomic Spectr. B, ed. W. Hanle and H. Kleinpoppen, p.1427(1979)

6. T.C. Griffith and G.R. Heyland, Phys.Rep. $\underline{39}$C, 169(1978)

7. A. Rich, Rev.Mod.Phys.$\underline{53}$, 127(1981)

8. G.P. Lepage, Phys.Rev. A$\underline{16}$, 863(1977); W.E. Caswell and G.P. Lepage, Phys.Rev. A$\underline{18}$, 810(1978) and A$\underline{20}$, 36(1979)

9. R. Barbieri and E. Remiddi, Nucl.Phys. B$\underline{141}$, 413(1978)

10. G.H. Bearman and A.P. Mills, Jr., Phys.Lett. $\underline{56}$A, 350(1976)

11. M.H. Yam, P.O. Egan, W.E. Friese, and V.W. Hughes, Phys.Rev. A$\underline{18}$, 350(1978)

12. D.W. Gidley and P.W. Zitzewitz, Phys.Lett. $\underline{69}$A, 97(1978)

13. T.C. Griffith, G.R. Heyland, K.S. Lines, and T.H. Twomey, J.Phys. B$\underline{11}$, L743(1978)

14. A.P. Mills, Jr., S. Berko, and K.F. Canter, Phys.Rev.Lett. $\underline{34}$, 1541(1975)

15. A.P. Mills, Jr., and L. Pfeiffer, Phys.Rev.Lett. $\underline{43}$, 1961(1979)

16. A.P. Mills, Jr., Appl.Phys.Lett. $\underline{35}$, 427(1979)

17. K.F. Canter, A.P. Mills, Jr., and S. Berko, Phys.Rev.Lett. $\underline{34}$ 177(1975)

18. A.I.Alekseev, Sov.Phys. JETP $\underline{34}$, 826(1958)

19. L.S. Vasilenko, V.P. Chebotaev and A.V. Shishaev, Sov.Phys.JETP Lett.$\underline{12}$, 113(1970)

20. M. Deutsch, Phys.Rev. $\underline{82}$, 455(1951)

21. S.L. Varghese, E.S. Ensberg, V.W. Hughes, and I. Lindgren, Phys.Lett. $\underline{49}$A, 415(1974)

22. H.S.W. Massey, E.H.S. Burhop, and H.B. Gilbody, "Electronic and Ionic Impact Phenomena", Vol.5 (Oxford Univ. Press) London, 1974

23. S.M. Curry, Phys.Rev. A$\underline{7}$, 447(1973);
    M.L. Lewis and V.W. Hughes, Phys.Rev. A$\underline{8}$, 2845(1973)

24. T. Fulton and P.C. Martin, Phys.Rev. $\underline{95}$, 811(1954)

25. M. Knudsen, Ann.Phys. (Leipzig) $\underline{48}$, 1113(1915)

26. V.S. Letokhov and V.G. Minogin, Sov.Phys.JETP $\underline{44}$, 70(1976)

27. T.W. Hänsch, S.A. Lee, R. Wallenstein, and C. Wieman,
    Phys.Rev.Lett. $\underline{34}$, 307(1975)

28. C. Wieman and T.W. Hänsch, Phys.Rev. A$\underline{22}$, 192(1980)

29. F.V. Kowalski, R.E. Teets, W. Demtröder, and A.L. Schawlow, J.Opt.
    Soc.Am. $\underline{68}$, 1611(1978)

30. J.E.M. Goldsmith, E.W. Weber, F.V. Kowalski, and A.L. Schawlow,
    Appl.Opt. $\underline{18}$, 1983(1979)

31. B. Cagnac, G. Grynberg, and E. Biraben, J.Physique $\underline{34}$, 845(1973)

32. Y. Gontier and M. Trahin, Phys.Rev. A$\underline{4}$, 1896(1971)

33. H.A. Bethe and E.E. Salpeter, "Quantum Mechanics of One- and
    Two-Electron Atoms"(Academic Press) New York, 1957

34. J.E.M. Goldsmith, E.W. Weber, and T.W. Hänsch, Phys.Rev.Lett $\underline{41}$,
    1525(1978)

35. M. Stobbe, Ann.Phys. $\underline{7}$, 661(1930)

36. D.W. Gidley, P.W. Zitzewitz, K.A. Marko, and A. Rich,
    Phys.Rev.Lett. $\underline{37}$, 729(1976)

37. A.P. Mills, Jr., Appl.Phys. $\underline{22}$, 273(1980)

38. W.H. Wing, Phys.Rev.Lett. $\underline{45}$, 631(1980)

39. G.P. Lepage, Atomic Physics 7 (Plenum Press, New York) 1981

40. E. Borie, "Quantum Electrodynamics of Bound Systems", in Proc. of
    the Symposium on Present Status and Aims of Quantum Electro-
    dynamics, eds. G. Gräff, E. Klempt, and G. Werth

41. R.A. Ferell, Phys.Rev. $\underline{84}$, 858(1951)

42. E.R. Cohen and B.N. Taylor, J.Phys.Chem.Ref.Data $\underline{2}$, 663(1973)

MUONIUM AND NEUTRAL MUONIC HELIUM

H. Orth

Physikalisches Institut der

Universität Heidelberg, W. Germany

1.   Introduction

Muonium (μ^+e^-) is the atom consisting of an electron and a
positive muon, and neutral muonic helium ($\alpha\mu^-e^-$) is the atom consisting
of a negative muon bound to a ^{4}He nucleus and a electron. These atoms
are isotopes of hydrogen since each contains one electron and a
positively charged muonic core. To study these simple atoms gives
information of the electromagnetic interactions of two different
leptons, testing the viewpoint that the muon behaves like a heavy
electron. Measurements of the atomic structure can be compared to
calculations very precisely and a value of the fine structure constant
α can be determined. In addition the properties of both the positive
and the negative muon such as the magnetic moment and the mass can be
accurately measured by independent experiments.

Up to the present time the only energy intervals that have been
measured are the hyperfine structure and Zeeman effect in the
electronic $1^2S_{1/2}$ ground state[1,2]. The method used for studying these
unstable atoms rely on the parity violation in the pion-muon-electron
decay sequence. The formation of the atom[3,4] with polarized muons from
pion decay results in unequal populations of ground state Zeeman
levels. This residual atomic polarization can be destroyed by induced
microwave magnetic resonance transitions which, in turn, can be
observed via the asymmetry in the angular distribution of the muon
decay electrons.

In this brief article the current status on muonium spectroscopy
with emphasis on recent developments will be summarized. The
experimental and theoretical progress of the muonic helium atom will be
reviewed. Future directions in this field of research will be
discussed.

2.   Muonium

Muonium is the simplest bound state of the positive muon and the electron. This purely leptonic atom is an extremely attractive testing ground for QED. Important advances in both theory and experiment of the hyperfine structure interval of muonium have been achieved recently. Further improvements are to be expected in the next few years.

2.1. Theoretical

The theory of the ground state splitting, $\Delta\nu$,[5] begins with the Fermi formula, which is the nonrelativistic expectation value of the contact interaction between the electron and the muon.

$$\Delta\nu_F = \frac{16}{3} \alpha^2 cR_\infty \mu_\mu/\mu_B \left(1 + \frac{m_e}{m_\mu}\right)^{-3} \tag{1}$$

The full theoretical expression is computed from the Bethe-Salpeter equation and gives modifications to the Fermi formula as an expansion in the fine structure constant α and the ratio $\rho = m_e/m_\mu$ of the electron to the muon mass

$$\Delta\nu_{th} = \Delta\nu_F \left(1 + f(\alpha,\rho)\right) \tag{2}$$

While $\Delta\nu_F$ contains only quantities to be determined experimentally the task of the theorist is to compute the term summarized by $f(\alpha,\rho)$. Using the conventional notation:

$$f(\alpha,\rho) = \frac{3}{2}\alpha^2 + a_e + \varepsilon_1 + \varepsilon_2 + \varepsilon_3 - \delta'_\mu + \tag{3}$$
$$+ \text{ higher order terms}$$

The term $3/2\alpha^2$ is the lowest order relativistic correction which arises from Dirac wavefunctions for the electron in a Coulomb field. Self energy corrections to the electron and vacuum polarization lead to:

$$\varepsilon_1 = \alpha^2 \left(\ln 2 - \frac{5}{2}\right)$$

$$\varepsilon_2 = - \frac{8\alpha^3}{3\pi} \ln\alpha \left(\ln\alpha - \ln 4 + \frac{281}{480}\right)$$

$$\varepsilon_3 = \frac{\alpha^3}{\pi} (18.4 \pm 5.0) \tag{4}$$

$$a_e = \frac{\alpha}{2\pi} - 0.328\ 478 \left(\frac{\alpha}{\pi}\right)^2 + 1.184(7) \left(\frac{\alpha}{\pi}\right)^3$$

Note that ε_3 has been evaluated only approximately. a_e is the electron anomaly. These radiative correction terms are also found in the hydrogen hyperfine structure. QED corrections to the muon are incorporated in the muon moment μ_μ. Most relevant to muonium is the relativistic recoil, denoted δ'_μ. In the case of the hydrogen the corresponding term contains proton structure functions. Whereas it is rigorously calculable for the muonium atom. Up to the order $\sigma(\alpha^2, \rho)$ there is the following expression:

$$\frac{1}{\rho} \cdot \delta'_\mu = -\frac{3\alpha}{\pi} (1-\rho^2)^{-1} \ln\rho + A\alpha^2 \ln\alpha (1+\rho)^{-2} - B\alpha^2 \ln\rho$$
$$+ C\left(\frac{\alpha}{\pi}\right)^2 (\ln\rho)^2 - D\left(\frac{\alpha}{\pi}\right)^2 \ln\rho + E\alpha^2 \tag{5}$$

Recently performed calculation[5,6,7] give: A=2, B=0 and C=+2; D and E still await to become evaluated.

The numerical evaluation of this lengthy expression for $f(\alpha,\rho)$ may be inserted in Equ. 2:

$$\Delta\nu_{th} = \Delta\nu_F [1 + (957.64(0.60) + 0.14D + 0.26E) \times 10^{-6}] \tag{6}$$

If the missing terms of the recoil effect, D and E, are neglected, theory contributes an error to $\Delta\nu_{th}$ of 0.6 ppm, which comes from the approximate value of ε_3. Despite the tedious work still to be done, all the terms of this order will eventually be calculated. The remaining error from higher oder terms such as $\rho\alpha^3 (\ln\alpha)^2 \approx 5 \times 10^{-8}$ then will be very small.

2.2. Ground State Energy Levels

The Zeeman split energy levels in an external magnetic field H are shown in Fig.1. They are eigenvalues of the spin dependent part of the Hamiltonian for muonium:

$$H = a \hat{S}_e \cdot \hat{S}_\mu + g_e \mu_B \hat{S}_e \cdot \hat{H} + g_{\mu+} \mu_B^\mu \hat{S}_\mu \cdot \hat{H} \tag{7}$$

The solution is given by the famous Breit-Rabi-equation.

$$\nu_{F=1/2\pm1/2, M_F} = -\frac{\Delta\nu}{4} + g_{\mu+} \mu_B^\mu M_F \cdot H/h$$
$$\pm \frac{\Delta\nu}{2} (1 + 2M_F X + X^2)^{1/2} \tag{8}$$

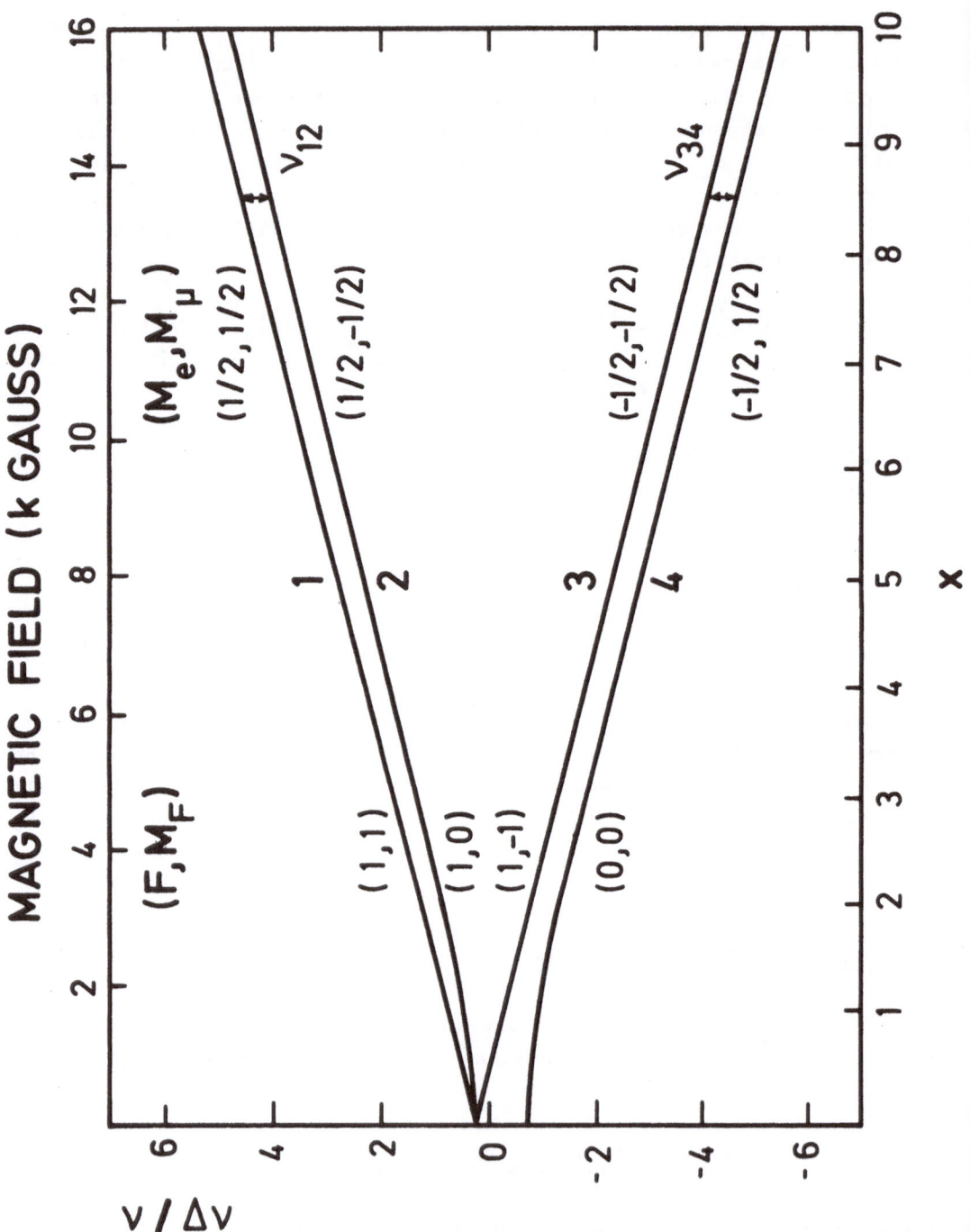

Fig.1: Diagram of the Zeeman split hyperfine structure in the ground
state of muonium. ν_{12} and ν_{34} are the muon spinflip transitions at 13.6 kG.

with $\qquad \Delta\nu = a/h \qquad$ and $\qquad X = (g_e - \rho g_{\mu+})\, \mu_B \cdot H/h\Delta\nu$

g_e and $g_{\mu+}$ are the bound state g-factors which can be expressed in terms of the free g-factors

$$g_e = 2(1 - \frac{\alpha^2}{3})(1 + a_e)$$

$$g_{\mu+} = -2(1 - \frac{\alpha^2}{3})(1 + a_\mu)$$

(9)

The Zeeman eigenstates, labelled by their (F, M_F) quantum numbers (or just by 1 to 4) may be written in terms of the muon and electron spin-eigenfunctions:

1: $\qquad (1,1) = \alpha_e \alpha_\mu$

2: $\qquad (1,0) = c\alpha_e \beta_\mu + S\beta_e \alpha_\mu$

3: $\qquad (1,-1) = \beta_e \beta_\mu$

4: $\qquad (0,0) = c\beta_e \alpha_\mu - S\alpha_e \beta_\mu$

(10)

$$c = \frac{1}{\sqrt{2}} (1+y)^{1/2} ; \qquad S = \frac{1}{\sqrt{2}}(1-y)^{1/2} ; \qquad y = x/\sqrt{1+x^2}$$

From these equations the occupance of Zeeman substates is determined, if the muon has been prepared in a definite spin state prior to muonium formation.

2.3. Experimental

Precision muonium experiments follow the main lines set forth by Hughes[3]. Polarized muons are stopped in a target medium which allows the muons to come to rest while forming muonium and spend their lifetime without depolarizing. The target chamber also serves as microwave cavity whose transverse magnetic field induces a resonant spin-flip transition. A resonance signal through the change in angular distribution of the decay positrons which accompanies this spin-flip is monitored. The signal may be traced as a function of the microwave frequency or of the external magnetic field, so that a resonance line is obtained. Depending on the applied experimental technique, the linewidth of the resonance curve is larger or smaller than the natural width given by the muon lifetime

$$\delta\nu = \frac{1}{\pi\tau} = 147 \text{ kHz.}$$

Transitions have been observed at several magnetic fields including very low fields of a few mGauss and strong fields up to 13.6 kGauss. Since the signal strength is proportional to the initial state population inequality which, in turn, depends on the external field (Equ.10), the different transitions exhibit their merits as well as their difficulties. In addition, by using refined experimental techniques, the precision of the experimental results can be greatly improved. With the advent of meson factories and newly developed high flux muon beams, accuracy of results is no longer limited by statistics only. Tab.1 is demonstrative of experimental improvements of the hfs interval $\Delta\nu$ during the last decade.

The earlier experiments[8,9] were all done by the conventional resonance technique, which involves taking the difference in counting rates between data with microwaves on and off. The muonium resonance line has the shape of a Lorentzian and the line width exceeds the natural width due to power broadening. A major advance in precision was made by using the Ramsey resonance technique[10] in zero magnetic field. It consists in applying two successive coherent microwave pulses separated by a time interval T and observing the change in the muon polarization, at times later than the end of the second pulse, caused by a relative phase shift of $\pm\pi/2$. By this method a resonance line narrower than the natural linewidth and a value of $\Delta\nu$ accurate to 0.4 ppm was obtained[11,12]. However, the split field technique has systematic dependences of the line shape caused by off-resonance cavity ringing and on-resonance phase shifts due to microwave pulsing. These problems are even more severe if the method is applied to transitions in an external magnetic field[13] where the resonance condition is fulfilled for only two levels.

At the LAMPF proton linac a zero field muonium resonance experiment has been performed using also the "old muonium" line narrowing technique[14]. There, magnetic transitions are induced in a relatively weak single oscillating field and decay positrons are selected from long lived muonium atoms. This technique is much less susceptible to systematics, yet the statistical power is comparable to the separated oscillating field technique.

Measurements of the Zeeman transitions in a strong magnetic field supply precise information on the muon magnetic moment μ_μ in addition to the hyperfine interval $\Delta\nu$. The general technique is to observe the two Zeeman transitions:

$(M_e, M_\mu) = (1/2, 1/2) \leftrightarrow (1/2, -1/2)$ and $(-1/2, -1/2) \leftrightarrow (-1/2, 1/2)$

at the same magnetic field. From the resonance frequencies $\nu_{12} = \nu_1 - \nu_2$ and $\nu_{34} = \nu_3 - \nu_4$ we obtain

$$\nu_{12} + \nu_{34} = \Delta\nu \tag{11}$$

$$\nu_{34} - \nu_{12} =: \Delta\mu = -2 g_{\mu+} \mu_B^\mu \cdot H/h + \Delta\nu[(1+x^2)^{1/2} - x]$$

$\Delta\mu$ is the spinflip frequency of the muon in a magnetic field plus a term which vanishes in the Paschen Back approximation $(x \gg 1)$. The magnetic moment is deduced from $\Delta\mu$ and comes out in units of the proton magnetic moment, because the magnetic field is measured by NMR. The precision in μ_μ/μ_p generally increases linearly with the applied magnetic field strength provided the accuracy of the field within the muon stopping distribution is sufficient.

At 11.3 kGauss $\Delta\mu$ is independent of the external field to first order $(\frac{\delta\Delta\mu}{\delta H} = 0$, "magic field"). Thus a measurement there significantly alleviates the homogeneity requirements which otherwise may seriously contribute to the experimental error in μ_μ. The price paid is the fact, that a high power tunable microwave system is required to scan the resonance lines via frequency. Recent muonium microwave magnetic resonance experiments have therefore been done at a slightly higher field of 13.6 kGauss[1].

Fig.2 shows the apparatus of the most recent experiment[15] at the LAMPF "surface" muon beam. This intense positive muon beam[16] (momentum 28 MeV/c, polarization 100 %) can be stopped in a 0.3 atm krypton gas target 20 cm in length and provides a production rate of about 10^6/s polarized muonium atoms. The thin scintillation counter S1 monitors the incoming muons. Energetic decay positrons traverse the moderator and are detected by the counters S2, S3. A central element in this experiment is the large solenoid which provides the magnetic field. It is homogeneous to a few ppm over the region of the cylindrical microwave cavity (r=10 cm, l=20 cm) and is stabilized to better than 1 ppm by NMR. Fig.3 shows a typical resonance curve obtained in a period of about three hours by varying the magnetic field. At each field point the microwave frequency is switched between the "upper" resonance transition ν_{12} and the "lower" resonance transition ν_{34}, so that the two signal points correspond to exactly the same magnetic field value. The data taking technique is of the conventional type in which the

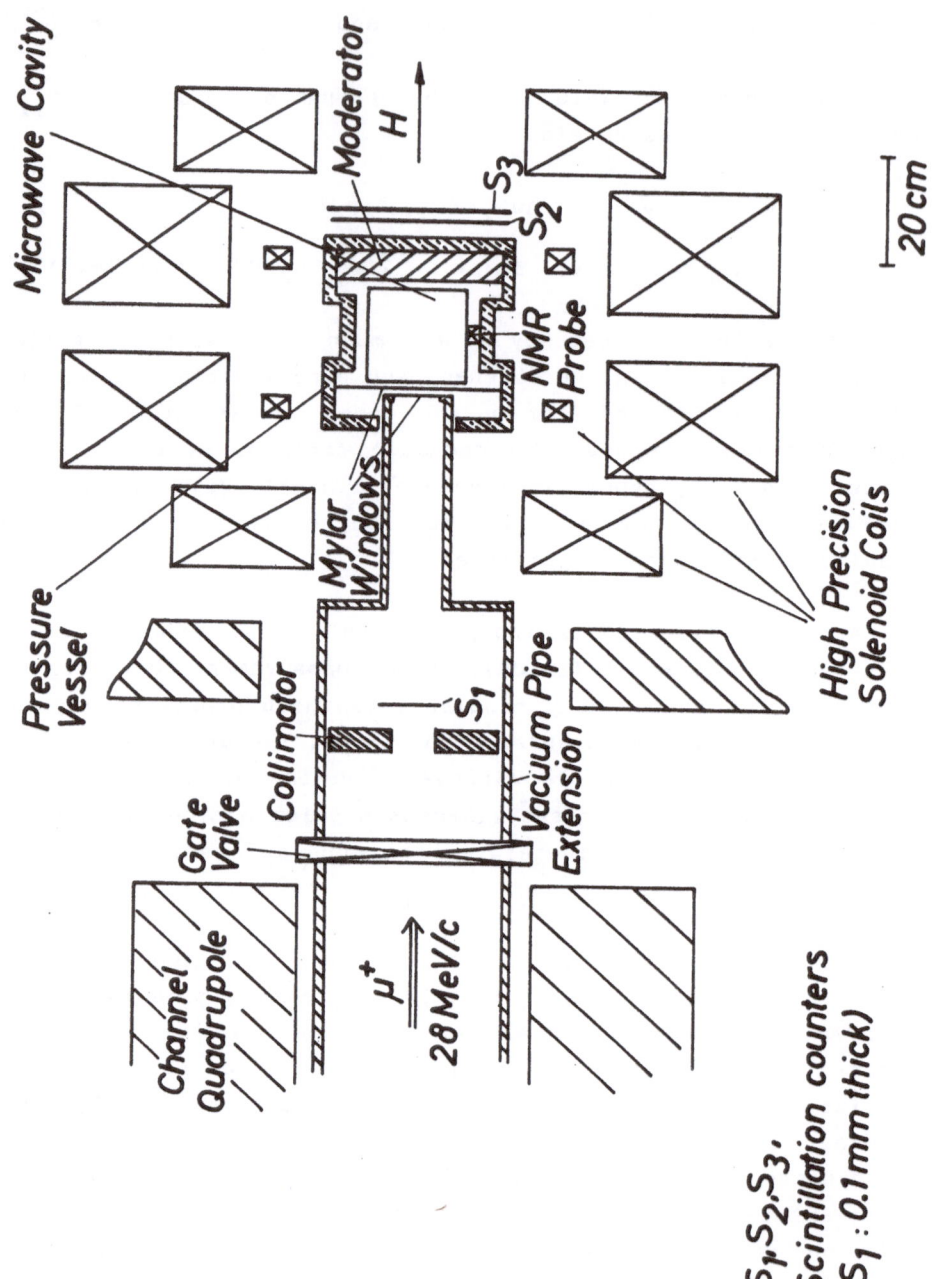

Fig.2: Schematic view of the strong field apparatus for muonium using surface muons.

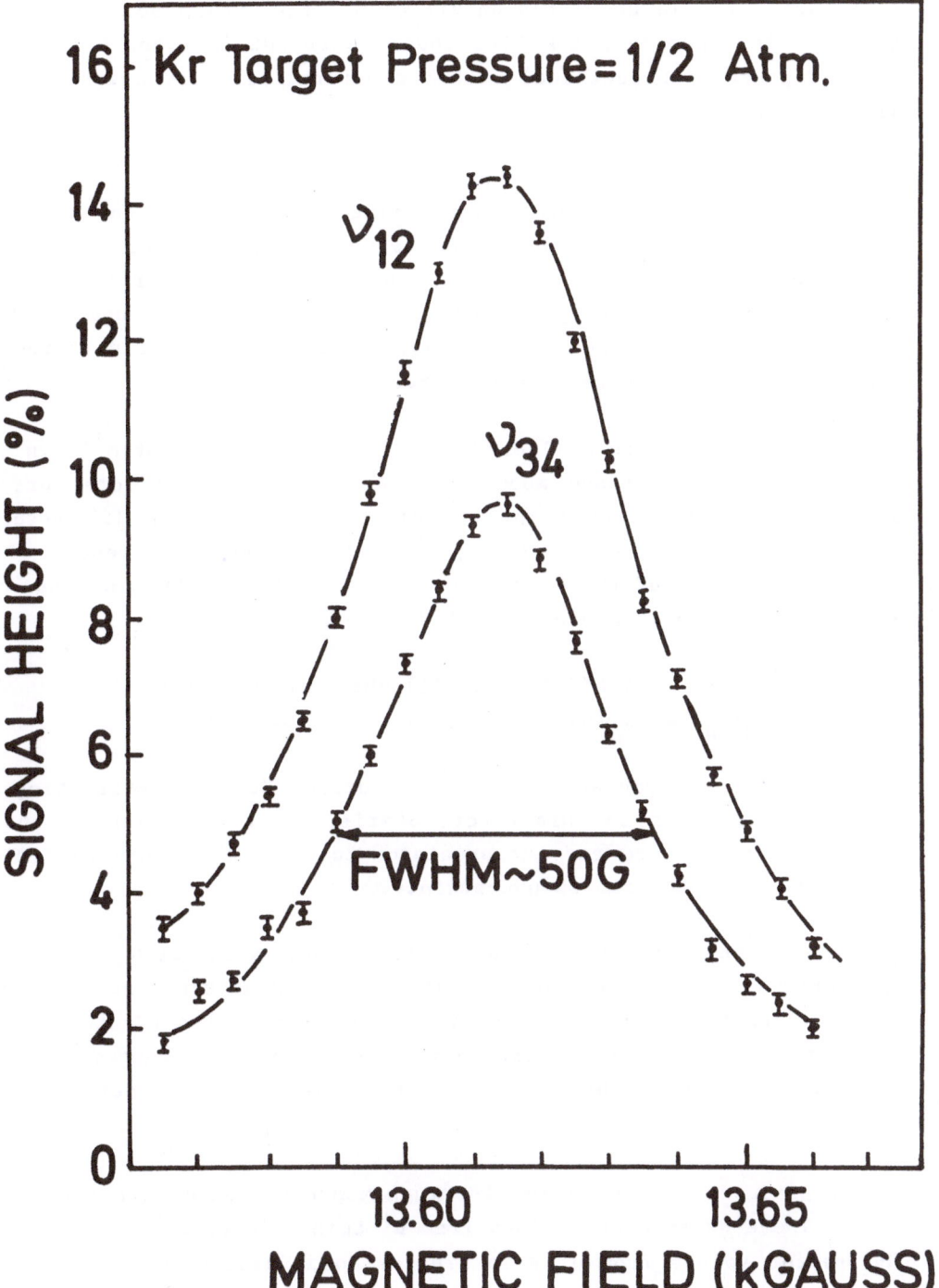

Fig.3: Muonium resonance curve for transitions ν_{12} and ν_{34} with fits to a Lorentzian line shape.

microwave power is modulated on-off with a few cycles per second repetition rate. Therefore the line shape is Lorentzian (apart from small
field dependent corrections) with the natural slightly power broadened
line width.

Inherently connected to precision spectroscopy of muonium in a
buffer gas is the fact, that the experimental Δν (and the electronic
g-factor) is affected by collisions with gas atoms leading to the so-
called density shift. This effect must be corrected for by an extrapolation to zero density (vacuum), using both linear and quadratic
terms. Therefore resonance curves have to be measured at different gas
densities at the same gas temperature.

In 1976 data were collected in 1.7 and 5.2 atm krypton[1]. In 1978,
the low momentum surface muon beam was available, and lower pressure
data in 0.5 and 1 atm krypton could be taken (about 150 resonance
curves of the type shown in Fig.3[15]). The preliminary result of the
combined analysis of these data for the hfs interval Δν and the ratio
μ_μ/μ_p of the muon to proton magnetic moments is:

$$\Delta\nu = 4\ 463\ 302.91\ (11)\ \text{kHz} \quad (0.025\ \text{ppm})$$
$$\mu_\mu/\mu_p = 3.183\ 344\ 78\ (96)\ \text{kHz} \quad (0.3\ \text{ppm})$$

One standard deviation errors are given. Counting statistics contribute about 60 % to this error. Statistical fluctuation of the experimental parameters and the precision to which the magnetic field is
known add up to the quoted uncertainty.

Tab.1 shows that this value of Δν agrees well with the results
from all the earlier experiments. The precision attained now is 5 parts
in 10^{-4} from the natural line width, which constitutes a natural boundary for the experimental resolution. Thus it will become extremely
difficult to surpass this precision of Δν by another order of magnitude.

Experimental results for the muon magnetic moment are compared in
Tab.2. There exist older values from μ^+ spin precession in water[17,18].
The recent values from μ^+ spin precission in liquid bromine[19,20] have
about the same accuracy as the new value from muonium. Agreement is
excellent and confirms that systematic effects causing difficulties for
either of these different experiments, are well understood.

Tab.1: Experimental result of the muonium hyperfine structure.
The following abbreviations are used:

CRT: conventional resonance technique
SOFT: separated oscillating field technique
OMT: "old muonium" technique

Method	Magnetic field	$\Delta\nu$	Error	Ref
CRT	H = 3 G, H < 10 mG	4 463 308(11)	2.5 ppm	9
CRT	H = 11 340 G "magic field"	4 463 313(18)	4.1 ppm	8
SOFT	H < 10 mG	4 463 304.0(1.8)	0.4 ppm	11,12
OMT, SOFT	H < 10 mG	4 463 302.2(1.4)	0.3 ppm	14
CRT	H = 13.6 kG	4 463 302.35(52)	0.12 ppm	1
CRT	H = 13.6 kG	4 463 302.91(11)	0.03 ppm	21

Tab.2: Precision measurements of the muon to proton magnetic moment ratio.

Method	Magnetic field	μ_{μ^+}/μ_p	Error	Ref
Precession of μ^+ in H_2O	H = 11 kG	3.183 330(44)	14 ppm	17
Presession of μ^+ in H_2O	H = 7.5 kG	3.183 346 7(82)	2.6 ppm	18
Muonium	H = 13.6 kG	3.183 342 0(44)	1.4 ppm	1
Precession of μ^+ in Br	H = 7.4 kG	3.183 344 8(29)	0.9 ppm	19
Muonium	H = 13.6 kG	3.183 347 8(26)	0.8 ppm	15
Precession of μ^+ in Br	H = 7.4 kG	3.183 344 1(17)	0.53 ppm	20
Muonium	H = 13.6 kG	3.183 344 78(96)	0.30 ppm	21
average	—	3.183 344 61(83)	0.26 ppm	

The averaged magnetic moment ratio μ_μ/μ_p from muonium and μ^+e^-Br spin precession, which has an error of only 0.26 parts in 10^{-6}, can be combined with the ratios μ_e/μ_p = 658.210 6880 (66) and g_μ/g_e = 1.00000626 (1)[29] to obtain the muon to electron mass ratio

$$\frac{m_\mu}{m_e} = (\mu_\mu/\mu_p)^{-1}(\mu_e/\mu_p)(g_\mu/g_e) = 206.768\ 317\ (60) \qquad (12)$$

If modern theoretical physics may eventually comprehend the nature of the muon and what causes the existence of this "heavy electron", the very precisely measured mass ratio has to be explained.

2.4. Conclusions and Outlook

The comparison of the theoretical prediction of $\Delta\nu$ with experiments is shown in Tab.3. A theoretical $\Delta\nu_{th}$ is calculated using Equ.2. The difference between $\Delta\nu_{th}$ and $\Delta\nu_{exp}$ is well within the errors, in which the contribution from theoretical uncertainties dominates. This agreement provides one of the most sensitive tests of quantum electrodynamics. It may be exploited to determine the fine structure constant from muonium, which to compare with the values from the helium fine structure[22], the anomalous magnetic moment of the electron[23] and with the ac-Josephson effect[24] exhibits the marvelous consistency within distinct branches of physics: the elaborate atomic physics calculations performed for the helium atom, the theory of superconductivity and quantum electrodynamics.

Further improvement of α from the hfs of muonium is definitely very promising. After completion of the calculation of all the radiative and recoil terms the remaining theoretical error in $\Delta\nu$ will be about 200 Hz, which is of the order of the experimental uncertainty of this quantity. It is an order of magnitude smaller than the error in μ_μ/μ_p presently contributing an uncertainty of 0.1 ppm to α. Thus the present accuracy may become the limiting number for the comparison of the muonium hfs with the theory.

For future experimental improvements of the muon magnetic moment, muonium seems to be most appropriate. An experiment ultimately may be performed at a magnetic field of about 160 kG, where a crossing of the "upper" Zeeman levels occures. Using muons polarized transverse to the momentum (spinrotator), the states (M_e,M_μ) = (1/2,+1/2) and (1/2,-1/2) can be populated coherently. The level crossing may be observed via a redistribution of the spacial asymmetry of decay positrons as a function of the external field near the crossing point.

TAb.3: Comparison of the muonium hfs interval with the theoretical prediction, and determination of the fine structure constant.

$$\Delta\nu = \alpha^2 \, \mu_\mu/\mu_p \; Q\big(1 + f(\alpha,\rho)\big)$$

$$Q = \frac{16}{3} cR_\infty \, \mu_p/\mu_B \, (1+\rho)^{-3}$$

$$Q = 2.630\,426\,58(3) \times 10^{13} \, s^{-1}$$

$$\alpha^{-1} = 137.035\,963(15)$$

$$\mu_\mu/\mu_p = 3.183\,344\,61(83)$$

$$\rho^{-1} = 206.768\,317(60)$$

$$f(\alpha,\rho) = 957.64(60) \times 10^{-6}$$

a.) **comparison experiment–theory**

$$\Delta\nu_{th} = 4\,463\,303.6(3.1) \; kHz$$

$$\Delta\nu_{exp} = 4\,463\,302.91(11) \; kHz$$

$$\Delta\nu_{th} - \Delta\nu_{exp} = 0.7(3.1) \; kHz$$

b.) **determination of α**

$$\Delta\nu_{exp} = \alpha^2 \, 8.381\,573\,12(59) \times 10^{10} \; kHz$$

$$\alpha^{-1} = 137.035\,973(48)$$

Tab.4: The fine structure constant; today's most precise determinations from different experiments.

Experiment	α^{-1}	ppm	Ref
helium fine structure	137.036 13(11)	0.8	22
ac–Josephson	137.035 963(15)	0.11	24
(g-2)-electron	137.036 006(11)	0.08	23
muonium	137.035 973(48)	0.3	21

A more realistic experiment would not really differ from the one described in chapter 2.2, but it would involve considerably higher magnetic field strength than 13.6 kG. It requires a pulsed muon beam and a superconducting solenoid of say 70 kGauss, with excellent stability and homogeneity of the magnetic field. At SIN a low momentum muon beam with a beam structure of 1 μs on and 10 μs off ideal to that sort of experiment is being planned. Building a solenoid adequate to this experiment is feasable with present day's technology. Hence an improvement of the experimental precision in μ_μ/μ_p by a factor of 10 is possible and may be achieved in the next few years.

Spectroscopy of muonium is presently restricted to experiments in the ground state. For precision QED tests a measurement of the Lambshift in the n=2 state would be particulary interesting. For this purpose muonium formation in the metastable 2S state is required which is impossible within a buffer gas, because muonium excited states are rapidly quenched due to collisions. Therefore the task of forming muonium in a vacuum-like environment has attracted a great deal of experimental effort[25,26,27,)]. Work has been concentrated on forming thermal muonium evaporating from metal foil targets in which muons have been stopped. But recent experiments at SIN[26)] and LAMPF[27)] must be interpreted, that thermal muonium production is very unlikely by such a mechanism. It seems much more promising to form energetic muonium from foils through which a low energy muon beam has passed. If muonium eventually can be isolated in a vacuum, a whole class of new exciting muonium spectroscopy will get within reach of the experimentalist.

3.   Neutral Muonic Helium

Neutral muonic helium $\alpha\mu^-e^-$, is a helium atom in which one of the electrons is replaced by a negative muon. The heavier muon orbits the helium nucleus in a hydrogen-like 1S state with energy and dimensions scaled by the muon reduced mass, so that from an atomic viewpoint the muon is tighly confined to the nucleus. The $(\alpha\mu^-)^+$ system then appears as a pseudonucleus with one unit of charge and mass $M=m_\alpha+m_\mu$, and with spin and magnetic moment equal to that of the muon. The remainung electron occupies a normal hydrogenic orbital about the $\alpha\mu^-$ pseudonucleus and the total atom has a gross structure of a hydrogen isotope of mass 4.11 amu.

The $\alpha\mu^-e^-$ is the simplest system for observing the electromagnetic interactions of the bound electron, including QED effects, in a muonic atom. In term of quantum mechanics it is an electromagnetic three-body bound state without exchange interactions.

The most accessible quantity for precision measurement in muonic helium is the hfs interval, $\Delta\nu$, in the ground state. It is expected to be similar in magnitude to that of muonium but inverted because of the negative moment of μ^- (compare Fig.1 with Fig.4). A precision determination of $\Delta\nu$ provides a very sensitive measurement of the specific μ^--e^- interaction in this atom. In particular, this is the first case where the Fermi contact interaction is precisely tested for two particles of like charges.

3.1 Theoretical

The theoretical value of $\Delta\nu$ for $\alpha\mu^-e^-$ can be written analogous to Equ.2

$$\Delta\nu = \Delta\nu_0 \left(1 + g(\alpha,\rho)\right) \tag{13}$$

in which $g(\alpha,\rho)$ contains the relativistic and radiative corrections to the nonrelativistic expectation value of the Fermi contact interaction between the two leptons.

$$\Delta\nu_0 = \frac{32\pi}{3h} \mu_\mu\cdot\mu_B \int\psi^*(\vec{r}_\mu,\vec{r}_e)\delta^3(\vec{r}_\mu-\vec{r}_e)\psi(\vec{r}_\mu,\vec{r}_e)d^3\vec{r}_\mu d^3\vec{r}_e \tag{14}$$

$\psi(\vec{r}_\mu,\vec{r}_e)$ is the wavefunction of the atom in the ground state. The pseudonucleus picture suggest dividing up this intergral in two parts, a leading term $\Delta\nu_F$, given by the Fermi formula for a point nucleus of mass $M=m_\alpha+m_\mu$

$$\Delta\nu_0 = \Delta\nu_F (1+\delta) \tag{15}$$

$$\Delta\nu_F = \frac{16}{3} \alpha^2 cR_\infty \mu_\mu/\mu_B \left(1 + \frac{m_e}{M}\right)^{-3} \tag{16}$$

and a correction term δ, which contains static as well as dynamic contributions associated with the finite size of the pseudonucleus. Although this pseudonucleus is quite small by atomic standards, it is nonetheless large compared to nuclear dimensions. In addition, the hfs interaction can be thought of being a selective filter for correlation effects between the electron and muon due to the delta function in Equ.14. Therefore the dynamic corrections exceed the static corrections

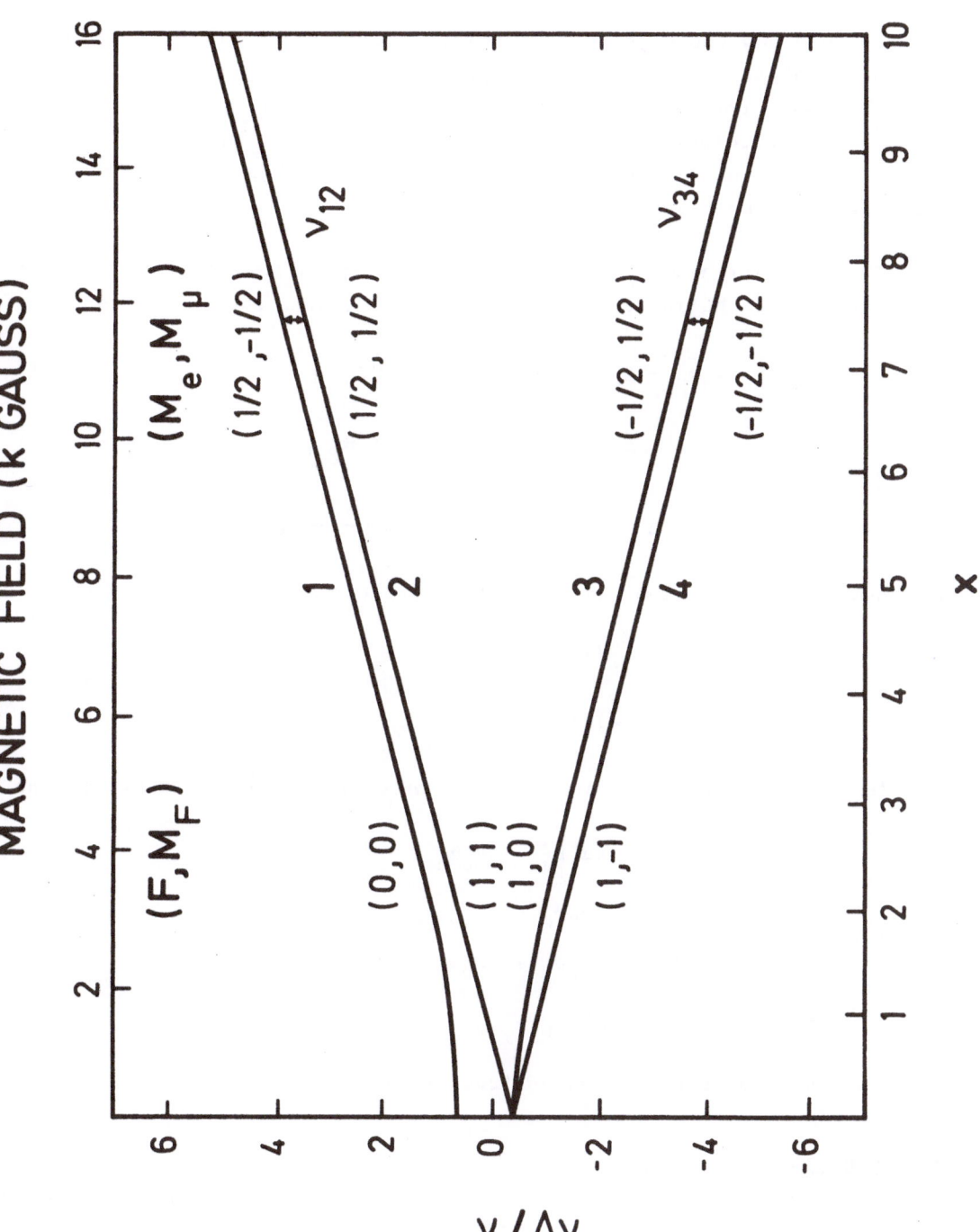

Fig.4: Diagram of the Zeeman split hyperfine structure in the ground state of muonic helium. The arrows at zero field and at 11.5 kG show the observed resonance transitions.

in magnitude and the problem of precision calculation of $\Delta \nu_0$ is much more difficult than has been anticipated[28,29]. Evaluation of $\Delta \nu$ has recently been done by a second-order perturbation approach[30] by a variational calculation[31] and by a Born-Oppenheimer approximation[32]. Values are given in Tab.4. The estimated error bars arise from the atomic physics calculation of $\Delta \nu$.

The formulas for the Zeeman energy levels in an external magnetic field for muonic helium are very similar to Equs. 7 and 10 for muonium.

$$H = a \hat{\vec{S}}_e \hat{\vec{S}}_\mu + g_e \mu_B \hat{\vec{S}}_e \hat{\vec{H}} + g_{\mu^-} \mu_B^\mu \hat{\vec{S}}_\mu \hat{\vec{H}} \tag{17}$$

$$\nu_{F=\frac{1}{2}\pm\frac{1}{2},M_F} = \frac{\Delta \nu}{4} + g_{\mu^-} \mu_B^\mu M_F H/h \mp \frac{\Delta \nu}{2} \left(1 - 2M_F x + x^2\right)^{1/2} \tag{18}$$

$$a = -h\Delta\nu \; ; \qquad x = \frac{(g_e - \rho g_{\mu^-})\mu_B H}{h\Delta\nu}$$

The approximate relation with the free-particle g-factors is given by

$$g_e = 2\left(1 - \frac{\alpha^2}{3}\right)\left(1 + a_e\right)$$

$$g_{\mu^-} = 2\left(1 - \frac{5\alpha^2}{3}\right)\left(1 + a_\mu\right) \tag{19}$$

and Zeeman eigenfunctions of the atom in terms of spin eigenfunctions are

$$
\begin{array}{llll}
1: & (0,0) & = & c\alpha_e \beta_\mu - s\beta_e \alpha_\mu \\[4pt]
2: & (1,1) & = & \alpha_e \alpha_\mu \\[4pt]
3: & (1,0) & = & c\beta_e \alpha_\mu + s\alpha_e \beta_\mu \\[4pt]
4: & (1,-1) & = & \beta_e \beta_\mu \; ,
\end{array}
\tag{20}
$$

c and s have the same meaning as from Equ.13.

3.2. Experimental

Determination of $\Delta \nu$ by a microwave magnetic resonance experiment requires, that $\alpha \mu^- e^-$ can be formed with some residual polarization. Negative muons brought to rest in pure helium gas do not form the atom, but rather the muonic helium ion, $(\alpha \mu^-)^+$, in the ground state which is energetically incapable of acquiring an electron from pure helium. If a xenon impurity is added to helium, the xenon acts as an electron

donor allowing the $(\alpha\mu^-)^+$ to be neutralized. The formation and polarization of $\alpha\mu^-e^-$ has been reported some years ago[33] using the μSR
technique. There, the characteristic precession frequencies of $(\alpha\mu^-)^+$
(ν_L=13.6 kHz/G) in pure helium and of $\alpha\mu^-e^-$ (ν_L=1.41 MHz/G) in helium
with a 2 % admixture of xenon was observed in the time distribution of
the decay electrons at the SREL polarized μ⁻ beam. Only about 2 to 3%
of the initial muon polarization is transferred into the atom which is
considerably smaller than the 9% expected from quite general arguements
of the μ-cascade and charge transfer process.

With regard to a microwave resonance experiment this residual
polarization is about 20 times smaller than it is in the analogous
muonium atom. Hence, the relatively small signal height renders the
experiment more difficult, if not an efficient repolarization method
for the atom were discovered.

It is very unlikely that the cascade depolarization can be
suppressed. However the mechanisme for the additional loss of muon
polarization is presently not understood, so there may be a chance to
circumvent it by an ingenious experimental technique. At present
repolarization using spin exchange with optically pumped alkaline atoms
is not feasable, because the target vessel is rather voluminous and a
helium pressure of a few atmospheres is necessary for copious muonic
helium formation.

Hfs transitions in the ground state of $\alpha\mu^-e^-$ have first been
observed in a microwave magnetic resonance experiment at zero magnetic
field[2]. The experiment was performed at SIN using the polarized μ⁻
beam (polarization 70%, momentum 55 MeV/c) at the μE4 channel. A
schematic diagram of the experimental set-up is shown in Fig.5. Muons
stopped in the gas within the microwave cavity are signalled by plastic
scintillation counters ($\mu_s = S \cdot M \cdot \bar{F}$). Decay electrons ($e_F, e_B$) are
identified by two scintillator telescopes (F,B) located forward and
backward with respect to the beam direction. Cavity and pressure vessel
are fabricated from high Z materials in which nuclear capture is more
likely than muon decay. Therefore background from muons stopped outside
the gas is efficiently reduced. Rates at 70 μA primary proton beam
typically are:

$$\mu_s(gas) = 2\times10^4 \ s^{-1} \ , \quad e_F = 1.5\times10^3 \ s^{-1} \ , \quad e_B = 1.2\times10^3 \ s^{-1}.$$

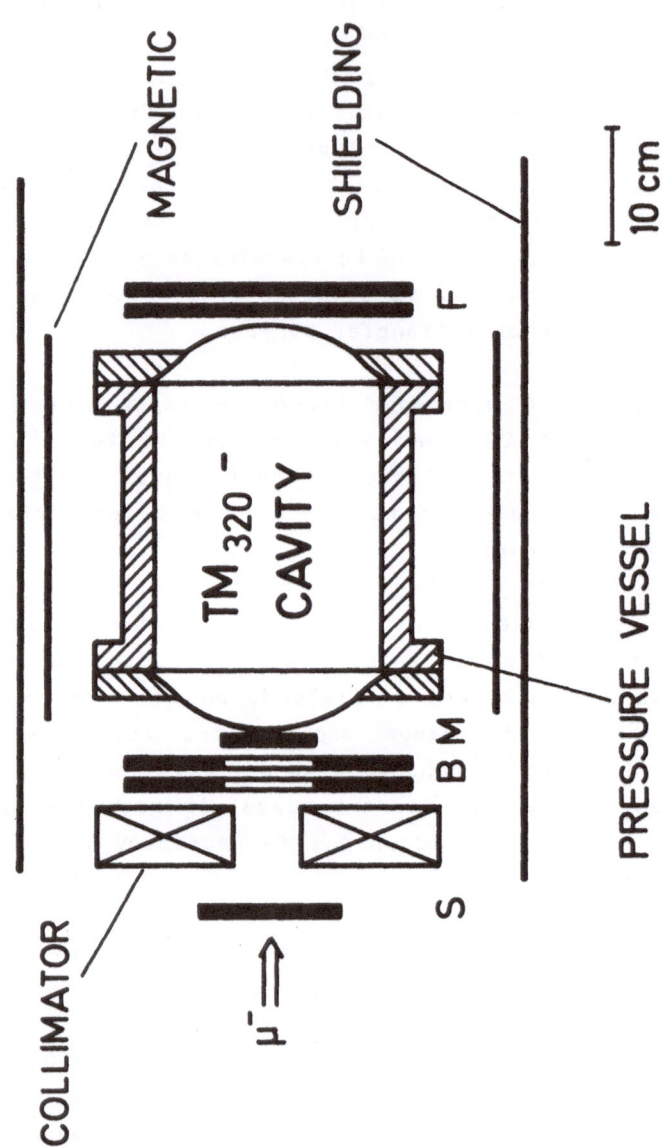

Fig.5: Schematic view of the set-up used to observe microwave magnetic
resonance transition in αμ⁻e⁻ at zero magnetic field.

Fig.6 shows the resonance signals as a function of the microwave frequency. It is obtained using the conventional data taking technique similar to the one of muonium.

The experimental result for the center of the resonance line is 4465.216 (56) MHz. In order to determine the vacuum value of the $\alpha\mu^-e^-$ hfs, a pressure shift extrapolation has to be applied to the resonance frequency. In that case this correction, of -0.264(17) MHz, could be measured with sufficient accuracy from muonium hfs transitions, which have been observed using exactly the same target gas. With the first experimental value for the $\alpha\mu^-e^-$ ground state hfs, $\Delta\nu$ = 4464.95(6) MHz it became apparent that the predictions of earlier calculation of this atomic structure had insufficient accuracy. There is good agreement with the most recent calculations within the large theoretical uncertainties (see Tab.5).

Observation of Zeeman hfs transitions ν_{12} and ν_{34} in this atom at strong magnetic field has been reported from LAMPF[34]. The experimental method is similar to that used in a strong field muonium experiment. The two spin-flip transitions can be combined (Equ.18) to determine the hfs interval and the magnetic moment of the negative muon.

Fig.7 shows a typical resonance line obtained by varying the magnetic field. Data are taken with gas pressures of 5 and 15 atm He and 1.5% Xe. Assuming a linear dependence of the hfs on density the extrapolated value for the free atom is: $\Delta\nu$ = 4465.001(40) MHz. If instead this extrapolation is determined from other hfs measurements on hydrogen[35,36] isotopes in helium and xenon: $\Delta\nu$ = 4464.974(17) MHz.

These results are based on apreliminary analysis of the data[34]. The value of $\Delta\nu$ from the strong field experiment agrees well with that from the zero field measurement and has a higher precision. From a combined analysis of both, the low and high field experiments, a determination of the magnetic moment of the negative muon with a precision of about 50 ppm can be expected.

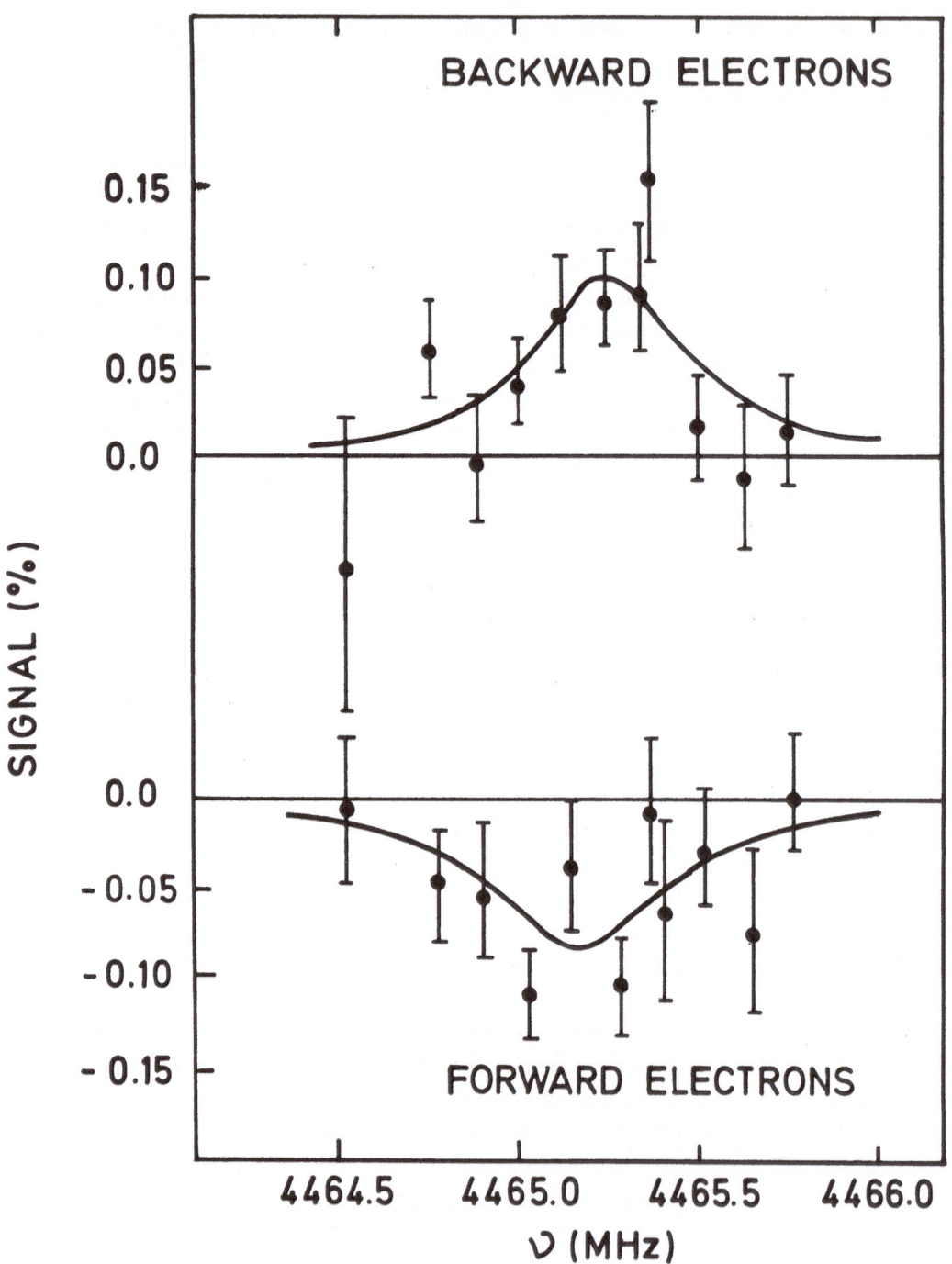

Fig.6: Resonance curves for the $\Delta F=1$, $\Delta M_F=\pm1$ hfs transition in $\alpha\mu^-e^-$, simultaneously observed in the backward (upper graph) and forward (lower graph) electron telescopes as a function of the microwave frequency.

Tab.5: Theoretical results of hfs interval calculations of $\alpha\mu^- e^-$ and comparison with values from experiments.

Method	$\Delta\nu(\alpha\mu^- e^-)$ (MHz)	Ref
hydrogenic wavefunctions with Bohr-Weisskopf correction	4511	28
Variational calculation (35 terms)	4494.14	29
Perturbational calculation	4462.6(3.0)	30
Variational calculation (496 terms)	4465.1(1.0)	31
Born-Oppenheimer approximation	4460	32
Zero field experiment (SIN) (density correction from $\mu^+ e^-$)	4464.95(6)	2
strong field experiment	density correction from $\alpha\mu^- e^-$	4465.001(40)	34
(LAMPF)	density correction from hydrogen	4464.974(17)	34

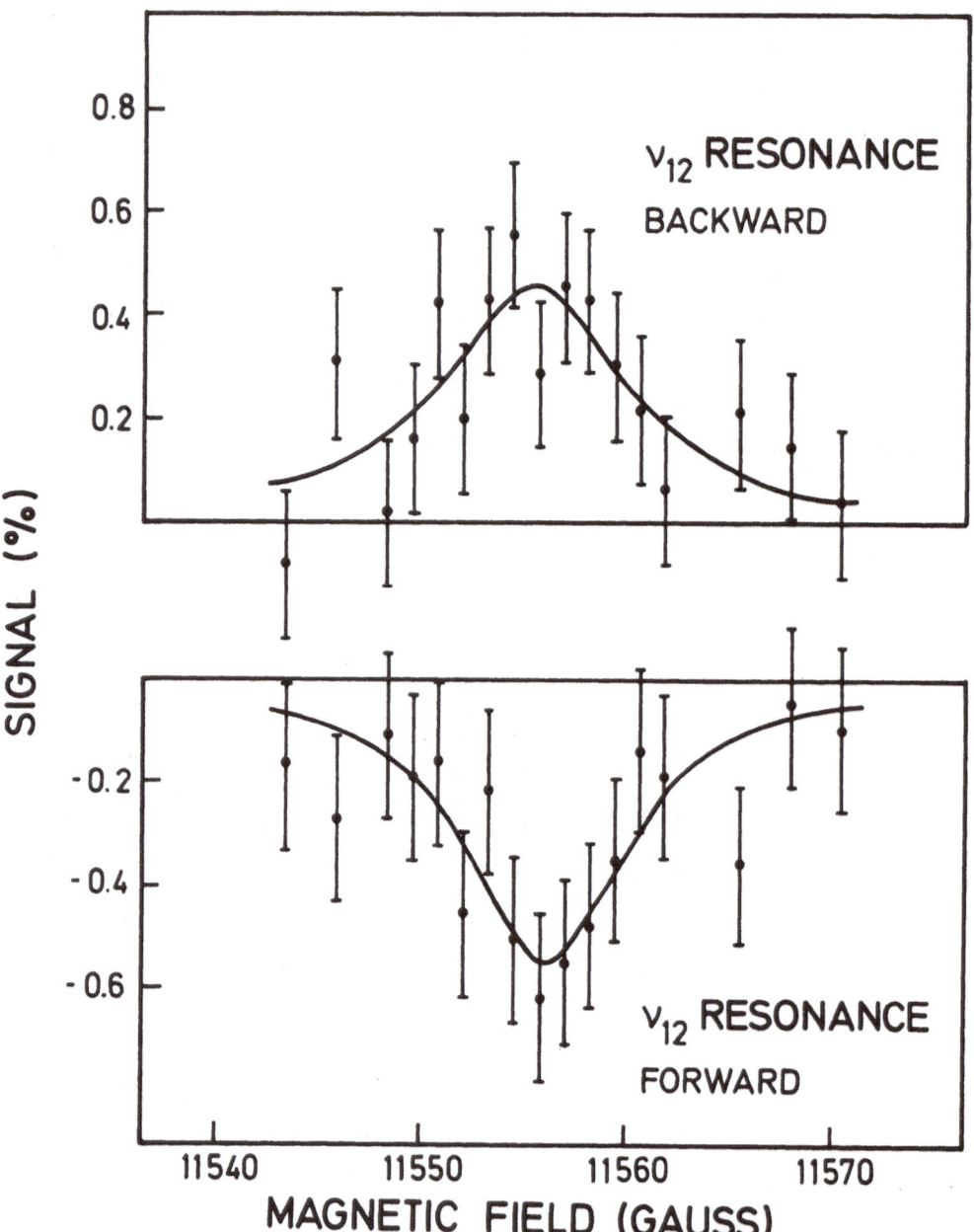

Fig.7: Resonance curves for the $(M_e, M_\mu) = (1/2, 1/2) \leftrightarrow (1/2, -1/2)$ Zeeman hfs transition of $\alpha\mu^- e^-$ with fitted theoretical line shape.

3.3. Conclusions

Measurements of the $\alpha\mu^-e^-$ hfs interval primarily test basic atomic structure calculations in which the major theoretical difficulty is the treatment of the three-body wavefunction in the nonrelativistic approximation. Apart from the g-factor anomalies of the electron, QED corrections in Equ.13 produce a -100 ppm shift on $\Delta\nu$[31,37]. Presently this contribution is masked by the error in the lowest order term $\Delta\nu_o$. Better calculations are very desirable since it is interesting to consider, what can be learned about the negative muon-electron interaction by studying this atom.

Another viewpoint is to compare the muonic helium hfs with the muonium hfs and look at the ratio $R = \Delta\nu(\alpha\mu^-e^-)/\Delta\nu(\mu^+e^-)$. There are no fundamental limitations to improve the measurement of $\Delta\nu$ ($\alpha\mu^-e^-$) by more than an order of magnitude. Thus R may be determined very accurately by the experiment since $\Delta\nu(\mu^+e^-)$ is already measured. Assuming the magnetic moments of the negative and positive muons to be the same, R contains only the atomic structure, the reduced mass factors and QED calculations

$$ R = (\frac{1 + m_e/m_\mu}{1 + m_e/M})^3 \ \frac{(1+\delta)(1+g(\alpha,\rho))}{1+f(\alpha,\rho)} \tag{21} $$

At least in principle the atomic structure can be known to very high precision. Then the measured value of R can be used as a test of QED calculations which, to first order, is in dependent of the fine structure constant.

Acknowledgement

The author wishes to thank K.P. Arnold, W. Beer, P.R. Bolton, P.O. Egan, C.K. Gardner, M. Gladisch, V.W. Hughes, W. Jacobs, D.C. Lu, G. zu Putlitz, F.G. Mariam, P.A. Souder, J. Vetter, W. Wahl and M. Wigand for their collective efforts making it possible to report on successful experiments. The financial support of the Bundesministerium für Forschung und Technologie is gratefully acknowledged.

References

1. D.E. Casperson, T.W. Crane, A.B. Denison, P.O. Egan, V.W. Hughes, F.G. Mariam, H. Orth, H.W. Reist, P.A. Souder, R.D. Stambaugh, P.A. Thompson, G. zu Putlitz, Phys. Rev. Lett. 38, 956 (1977); 1504 (1977)

2. H. Orth, K.P. Arnold, P.O. Egan, M. Gladisch, W. Jacobs, J. Vetter, W. Wahl, M. Wigand, V.W. Hughes, G. zu Putlitz, Phys. Rev. Lett. 45, 1483 (1980)

3. V.W. Hughes, Ann. Rev. Nucl. Sci. 16, 445 (1966)

4. P.A. Souder, T.W. Crane, V.W. Hughes, D.C. Lu, H. Orth, H.W. Reist, M.H. Yam, G. zu Putlitz, Phys. Rev. A22, 33 (1980)

5. G.T. Bodwin, D.R. Yennie, Phys. Reports 43, 267 (1978) and references therein.

6. G.T. Bodwin, D.R. Yennie, M.A. Gregorio, Phys. Rev. Lett. 41, 1088 (1978)

7. W.E. Caswell, G.P. Lepage, Phys. Rev. Lett. 41, 1092 (1978)

8. P.A. Thompson, P. Crane, T.W. Crane, J.J. Amato, V.W. Hughes, G. zu Putlitz, J.E. Rothenberg, Phys. Rev. A8, 86 (1973)

9. R.D. Ehrlich, H. Hofer, A. Magnon, D.Y. Stowell, R.A. Swanson, V.L. Telegdi, Phys. Rev. A5, 2357 (1972)

10. N.F. Ramsey, Phys. Rev. 78, 695 (1950)

11. D. Favart, P.M. McIntyre, D.Y. Stowell, V.L. Telegdi, R. DeVoe, R.A. Swanson, Phys. Rev. A8, 1195 (1973)

12. H.G.E. Kobrak, R.A. Swanson, D. Favart, W. Kells, A. Magnon, P.M. McIntyre, J. Roehrig, D.Y. Stowell, V.L. Telegdi, M. Eckhouse, Phys. Lett. 43B, 526 (1973)

13. W. Kells, P.M. McIntyre, J. Roehrig, V.L. Telegdi, H. Knapp, C.E. Kobrak, R.A. Swanson, Nuovo Cim. 35A, 289 (1976)

14. D.E. Casperson, T.W. Crane, V.W. Hughes, P.A. Souder, R.D. Stambaugh, P.A. Thompson, H. Orth, G. zu Putlitz, H.F. Kaspar, H.W. Reist, A.B. Denison, Phys. Lett. 59B, 397 (1975)

15. P.O. Egan, W. Beer, P.R. Bolton, C.K. Gardner, V.W. Hughes, D.C. Lu, F.G. Mariam, P.A. Souder, J. Vetter, H. Orth, G. zu Putlitz, U. Moser, Bull. Am. Phys. Soc. 25, 19 (1980)

16. H.W. Reist, D.E. Casperson, A.B. Denison, P.O. Egan, V.W. Hughes, F.G. Mariam, G. zu Putlitz, P.A. Souder, P.A. Thompson, J. Vetter, Nucl. Inst. Meth. 153, 61 (1978)

17. D.P. Hutchinson, F.L. Larsen, N.C. Schoen, D.I. Sober, A.S. Kanofsky, Phys. Rev. Lett. 24, 1254 (1970)

18. K.M. Crowe, J.F. Hague, J.E. Rothberg, A. Schenck, D.L. Williams, R.W. Williams, K.K. Young, Phys. Rev. D5, 2145 (1972)

19. M. Camani, F.N. Gygax, E. Klempt, W. Rüegg, A. Schenck, H. Schilling, R. Schulze, H. Wolf, Phys. Lett. 77B, 326 (1978)

20. Reanalysis of all the data from ref.19; value given to this symposium

21. Combined analysis of the data from ref. 1 and 15

22. W.E. Frieze, E.A. Hinds, V.W. Hughes, F.M.J. Pichanik, Phys. Lett. 78A, 322 (1980)

23. H. Dehmelt; value of α reported to this symposium

24. E.R. Williams, P.T. Olsen, Phys. Rev. Lett. 42, 1575 (1979) and references in there

25. B.A. Barnett, C.Y. Yang, P. Steinberg, G.B. Yodh, H.D. Orr, J.B. Caroll, M. Eckhouse, J.R. Crane, C.B. Spence, C.S. Hsieh, Phys. Rev. A15, 2246 (1977)

26. K.P. Arnold, P.O. Egan, M. Gladisch, W. Jacobs, H. Orth, G. zu Putlitz, J. Vetter, W. Wahl, M. Wigand, D. Herlach, H. Metz, Verhandl. DPG 2, 552 (1979)

27. W. Beer, P.R. Bolton, P.O. Egan, V.W. Hughes, D.C. Lu, F.G. Mariam, P.A. Souder, J. Vetter, M. Gladisch, G. zu Putlitz, U. Moser, Bull. Am. Phys. Soc. 24, 675 (1979)

28. E.W. Otten, Z. Physik 225, 393 (1969)

29. V.W. Hughes, T. Kinoshita, in Muon Physics edited by V.W. Hughes and C.S.Wu (Academic, New York, 1977) Vol.I, p.11.

30. S.D. Lakdawala, P.J. Mohr, Phys. Rev. A22, 1572, (1980)

31. K.-N. Huang, V.W. Hughes, Phys. Rev. A20, 706 (1979); A21, 1071 (1980)

32. R.J. Drachman, Phys. Rev. A22, 1751 (1980)

33. P.A. Souder, D.E. Casperson, T.W. Crane, V.W. Hughes, D.C. Lu, H. Orth, H.W. Reist, M.H. Yam, G. zu Putlitz, Phys. Rev. Lett. 34, 1417 (1975)

34. C.K. Gardner, W. Beer, P. Bolton, B. Dichter, P.O. Egan, V.W. Hughes, D.C. Lu, F.G. Mariam, P.A. Souder, H. Orth, G. zu Putlitz, J. Vetter, Bull. Am. Phys. Soc. 25, 19 (1980)

35. E.M. Pipkin, R.H. Lambert, Phys. Rev. 127, 787 (1962)

36. E.S. Ensberg, C.L. Morgan, Phys. Lett. 28A, 106 (1968)

37. E.Borie, Z. Physik A291, 107 (1979)

EUROPEAN ORGANIZATION FOR NUCLEAR RESEARCH

VACUUM POLARIZATION IN MUONIC ATOMS

L. Tauscher[*]

Institute of Physics, University of Basle,

Basle, Switzerland

[*] Visitor at CERN, Geneva, Switzerland.

1. INTRODUCTION

In the course of this meeting it became obvious that the muonic atom is not
the only tool for testing vacuum polarization. Although the classical experiments
on QED, such as the g-2 experiments, test essentially graphs of type (a) in Fig. 1,
their very great accuracy also provides very valuable tests for the much smaller
vacuum polarization [type (b) in Fig. 1]. The same is true for the Lamb shift
measurement in hydrogen. Table 1 summarizes the accuracy reached in these different measurements[1-3]. Of course this way of looking at these experiments only
makes sense if it is assumed that all other contributions are known theoretically.

The muonic atom is, however, a tool for studying and testing vacuum polarization at stronger fields and higher momentum transfers. Furthermore, the vacuum
polarization is the dominant QED correction to the lowest-order Bohr (or Dirac)
energies. Moreover, the difficulties of the μHe Lamb shift[4] measurement, introduced by the finite size of the α-particle, may be overcome by choosing transitions that are much less dependent on the finite size contribution.

It is obvious that in order to test the vacuum polarization, X-ray transitions have to be selected where uncertainties from contributions other than the
vacuum polarization are minimized with respect to it. Such transitions are the
3-2 transitions in light elements ($Z \approx 13$), the 4-3 transitions in medium heavy
elements ($Z \approx 56$) and the 5-4 transitions in heavy elements ($Z \approx 82$).

In Table 2 the various corrections[5-8] contributing to a muonic X-ray transition energy are listed for He, Mg, Ba, and Pb.

It is apparent that the calculations have uncertainties which are comparable
to or larger than the experimental errors. The error due to the finite size in
the case of μHe amounts to 0.25% of the vacuum polarization. In the case of Mg
this error contribution is less than 0.02%, and in Pb it is of the same order of
magnitude. The "theoretical" uncertainties in Mg and Pb arise mostly from the
electron screening. This uncertainty is not so much due to the calculation of
the electron screening potential but stems rather from the badly known number of

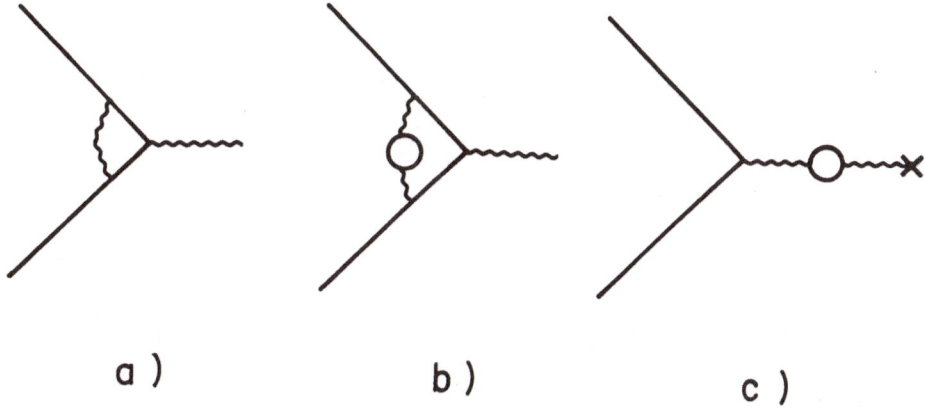

a)   b)   c)

Fig. 1   :   Basic QED corrections.

Table 1

Comparison of classical QED tests

Experiment	Exp. value	Vac. pol.	Test of vac. pol. (%)
(g-2) electron[1]	115 965.241 (4) $\times 10^{-8}$	9.4 $\times 10^{-8}$	0.02
(g-2) muon[2]	116 592.2 (9) $\times 10^{-8}$	585.6 $\times 10^{-8}$	0.15
Lamb shift (H)[3]	1057.893 (20) MHz	27.323 MHz	0.07

Table 2

Calculated contributions to various muonic X-ray transitions

	He $2s_{1/2}-2p_{3/2}$ (meV)	Mg $3d_{5/2}-2p_{3/2}$ (eV)	Ba $4f_{7/2}-3d_{5/2}$ (eV)	Pb $5g_{9/2}-4f_{5/2}$ (eV)
Point nucl.	145.6	56 038.9	431 652	429 343
Finite size	−288.9 (4.1)	−0.89 (3)	−55 (1)	−4 (1)
El. screening	−	−0.37 (30)	−17 (1)	−82 (3)
Lamb shift	−11.1	−0.15[a)]	−1.5	−0.9
Recoil	0.3	0.18	3.6	2.2
Nucl. pol.	3.1	0.07 (7)	7.9 (8)	4.5 (5)
Vac. pol.:				
$\alpha(\alpha Z)$	1666.1	177.5	2327.5	2105.0
$\alpha^2(\alpha Z)$	11.6	1.24	16.2	14.5
$\alpha(\alpha Z)^{3,5,\ldots}$	−	−0.15	−19	−42 (2)
Exp. accuracy	0.3	0.53	8	3.4

a) Contains the anomalous magnetic moment contribution.

Table 3

Crystal spectrometer results (in eV)

Transition	Exp. energy	Calc. energy	Vac. pol.
Mg $3d_{5/2}-2p_{3/2}$	56 216.3 ± 0.53	56 216.4 ± 0.30	177.5
$3d_{3/2}-2p_{1/2}$	56 392.7 ± 0.85	56 391.8 ± 0.30	179.3
Si $3d_{5/2}-2p_{3/2}$	76 617.6 ± 1.14	76 617.6 ± 0.43	273.2
$3d_{3/2}-2p_{1/2}$	76 941.4 ± 2.10	76 942.3 ± 0.43	276.8
P $3d_{5/2}-2p_{3/2}$	88 016.2 ± 2.37	88 015.9 ± 0.50	330.4
$3d_{3/2}-2p_{1/2}$	88 423.8 ± 8.2	88 443.2 ± 0.50	335.4

Table 4

Spectrometer characteristics

	CERN[8]	SIN[7]	SREL[11]
Type	2 × Ge(Li) planar	Ge(Li) coax.	Ge planar
Volume (cm³)	4.9/1.0	50	3.1
Resolution (keV)	1.49/1.31 at 412 keV	1.60 at 412 keV	0.87 at 316 keV

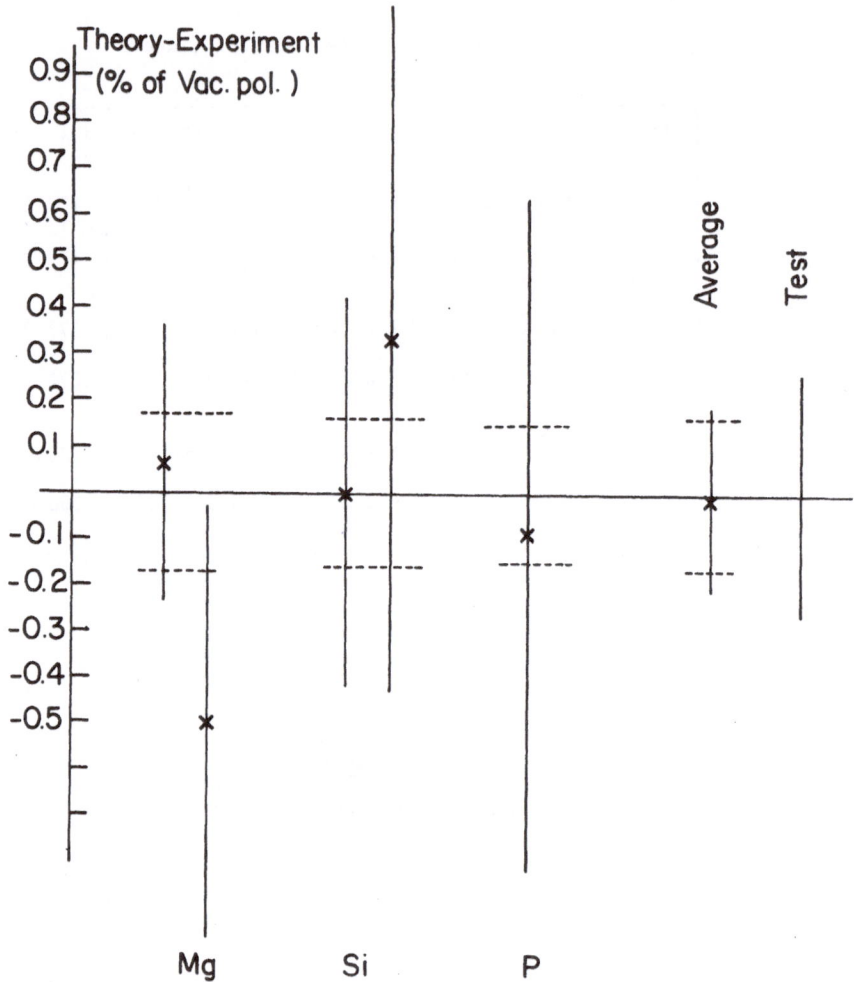

Fig. 2 : Deviation of the measured energies from the calculated ones according
to Ref. 6. Error bars are experimental errors, broken lines indicate
uncertainties from calculations.

s-electrons present when the muonic X-ray transition occurs. Table 2 also shows that the experimental accuracy for Pb is sufficient for it to become sensitive to those contributions which are due to the high fields [such as the ones in which the virtual e^+e^- cloud is polarized, terms of order $(\alpha Z)^3$]. These contributions are not important in the lighter atoms.

It should be noted that the energy of a muonic X-ray transition from one main quantum number to the next lower one is mainly due to the difference of Bohr binding energies, and the QED corrections amount to 0.5% or less of this transition energy. Thus a test of the vacuum polarization to 0.2% accuracy requires a precision of 10 ppm in the determination of the transition energy. This is an accuracy comparable to the one reached in the Lamb shift measurement on hydrogen or the g-2 measurement for the muon.

2. EXPERIMENTS

Two types of experiments were performed so far, using entirely different types of spectrometer [the laser technique of Zavattini[4] is not a subject of this talk].

At SIN, three elements were measured with the Fribourg *bent crystal spectrometer*. The target is placed in the muon channel, and stop rates of $\sim 0.5 \times 10^6$ s^{-1} may be obtained in a 25 mg/cm^2 thick target. The resolution of this spectrometer is of the order of 35 eV at 56 keV. Details may be found elsewhere[9]. Table 3 shows the measured transitions and their energies as well as the vacuum polarization contribution[6]. The calibration was done relative to γ-ray standards between 50 keV and 90 keV [10].

The accuracy obtained for the Mg $3d_{5/2}-2p_{3/2}$ transition is 9 ppm. The errors are mostly statistical (~ 5 ppm from line shape, 1-2 ppm from geometry, 7 ppm and more from statistics).

In Fig. 2 the deviation of the measured energies from the calculated ones $(E_{exp}-E_{calc})$ is displayed in units of percent of the vacuum polarization correction. The error bars are for the experimental errors only. The errors from the calculations are indicated by the broken lines.

The final result is that the crystal spectrometer measurements test the vacuum polarization to 0.2% (from the experimental errors alone).

The usual way of calculating the testing power of several measurements is to add the experimental and the "theoretical" uncertainties (linearly or quadratically) for each experiment and then deduce the weighted average of the deviation and its error. This is wrong since the "theoretical" errors are correlated and systematic. The correct way to determine the "theoretical" uncertainty for several transitions is to calculate the average of the individual "theoretical" uncertainties weighted with the corresponding experimental errors.

Combining the final experimental accuracy of 0.2% with the uncertainty thus obtained for the calculations of 0.17%, the vacuum polarization is tested to 0.26% at 1σ confidence.

Since 1976, three groups[7,8,11] have published precision experiments on heavy muonic atoms for testing the QED contributions. The measurements were done with *Ge diodes*. The spectrometer characteristics are shown in Table 4 (the CERN group used two diodes).

The results of the measurements are listed in Table 5. The energy calibration is with respect to an Au standard of 411,805.2 eV, whereas the final published value is 411,804.41 ± 0.15 eV [12]. Thus all experimental values might be too high by about 0.8 eV. (A re-adjustment seems, however, difficult in view of the unknown non-linearities in the different experiments.)

The experimental accuracy has to be considered carefully, since it is only in the CERN measurement that the statistical errors dominate (7-9 eV statistics, $\sim$ 6 eV background and calibration). The SIN group has a statistical error of 3-4 eV and, in linear addition, a 5 eV systematic error. The SREL group claims a statistical error, including background lines and calibration, of 0.7-3 eV, and a constant systematic error of 1.7 eV. Besides pure statistics, uncertainties in these measurements arise from unresolved background lines (X-rays of parallel transitions, nuclear γ-lines, etc.), from geometrical and count rate influences,

Table 5

Ge spectrometer results (in eV)

	SIN[7]	SREL[11]	CERN[8]	Calculation	Vac. pol.
Tl $5g_{9/2}-4f_{7/2}$	420 763 ± 8	420 757.3 ± 3.7		420 768 ± 6	2012 ± 2.2
$5g_{7/2}-4f_{5/2}$	426 865 ± 8			426 868 ± 6	2091 ± 2.2
Pb $5g_{9/2}-4f_{7/2}$	431 331 ± 8	431 327.6 ± 3.4	431 360 ± 11	431 337 ± 7	2079 ± 2.2
$5g_{7/2}-4f_{5/2}$	437 749 ± 8	437 749.4 ± 13.7	437 748 ± 12	437 750 ± 7	2163 ± 2.2
Ba $4f_{7/2}-3d_{5/2}$	433 897 ± 8	433 904.8 ± 9.6	433 926 ± 8	433 910 ± 3	2326 ± 1.4
$4f_{5/2}-3d_{3/2}$	441 358 ± 8	441 361.7 ± 5.1	441 374 ± 9	441 361 ± 4	2434 ± 1.4
Ce $4f_{7/2}-3d_{5/2}$	465 754 ± 8			465 748 ± 6	2547 ± 1.4
$4f_{5/2}-3d_{3/2}$	474 330 ± 8			474 329 ± 6	2671 ± 1.4

Table 6

Summary of QED test
(% of vacuum polarization)

	Crystal	Ge diodes
Exp. accuracy	0.20	0.07
Calc. uncertainty	0.17	0.25
Test	0.26	0.26

etc. All this is discussed carefully in the CERN and SIN experiments, whereas a detailed error discussion is not yet available for the SREL experiment.

The complexity of the experimental spectra is demonstrated in Fig. 3, where the relevant part of the CERN data is displayed.

In Fig. 4 the deviations of the measured energies from the calculated ones are displayed, again in units of percent of the QED contributions. The error bars correspond to the experimental errors, and the broken lines reflect the uncertainties in the calibration.

The measurements reach, in their average, an experimental accuracy of 0.07% for the test of the vacuum polarization. This accuracy is essentially due to the SREL points in Tl and Pb, for which no detailed error discussion is available yet. The weighted average of the uncertainties from the calculation is 0.25%. Thus the total test quality of these experiments is entirely governed by the "theoretical" uncertainties, and amounts to 0.26% at 1σ confidence.

3. SUMMARY AND OUTLOOK

It is obvious that the quality of the test of QED in muonic atoms depends not only on the experimental precision but also on the correctness of the calculated quantities. Unfortunately the latter is dominating in the case of muonic atoms, as is shown in Table 6.

The uncertainties in the calculation stem from various effects, such as electron screening, nuclear polarization, etc. As long as these effects cannot be fixed by separate experimental checks, any improvement in the experimental accuracy is meaningless.

There exist proposals to study the electron screening by carefully measuring the screening effect in higher orbits (where QED contributions are small), as proposed by the SIN crystal spectrometer group. Similar efforts have to be made for the heavy muonic atoms in order to study and to understand quantitatively the cascade processes. Efforts have also to be made, both experimentally and theoretically, to understand the nuclear polarization.

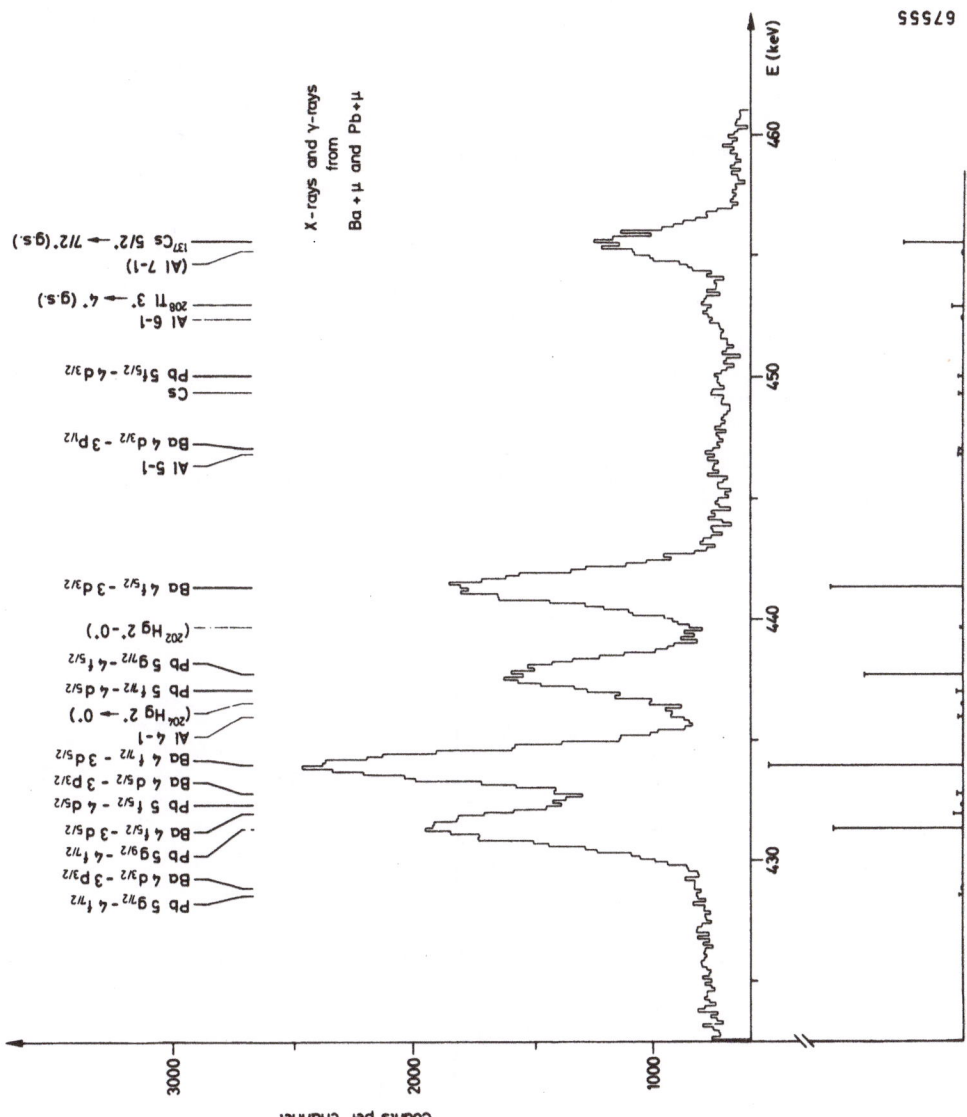

Fig. 3 : X-ray spectrum as obtained in Ref. 8. Background lines are indicated at the bottom with their relative intensities.

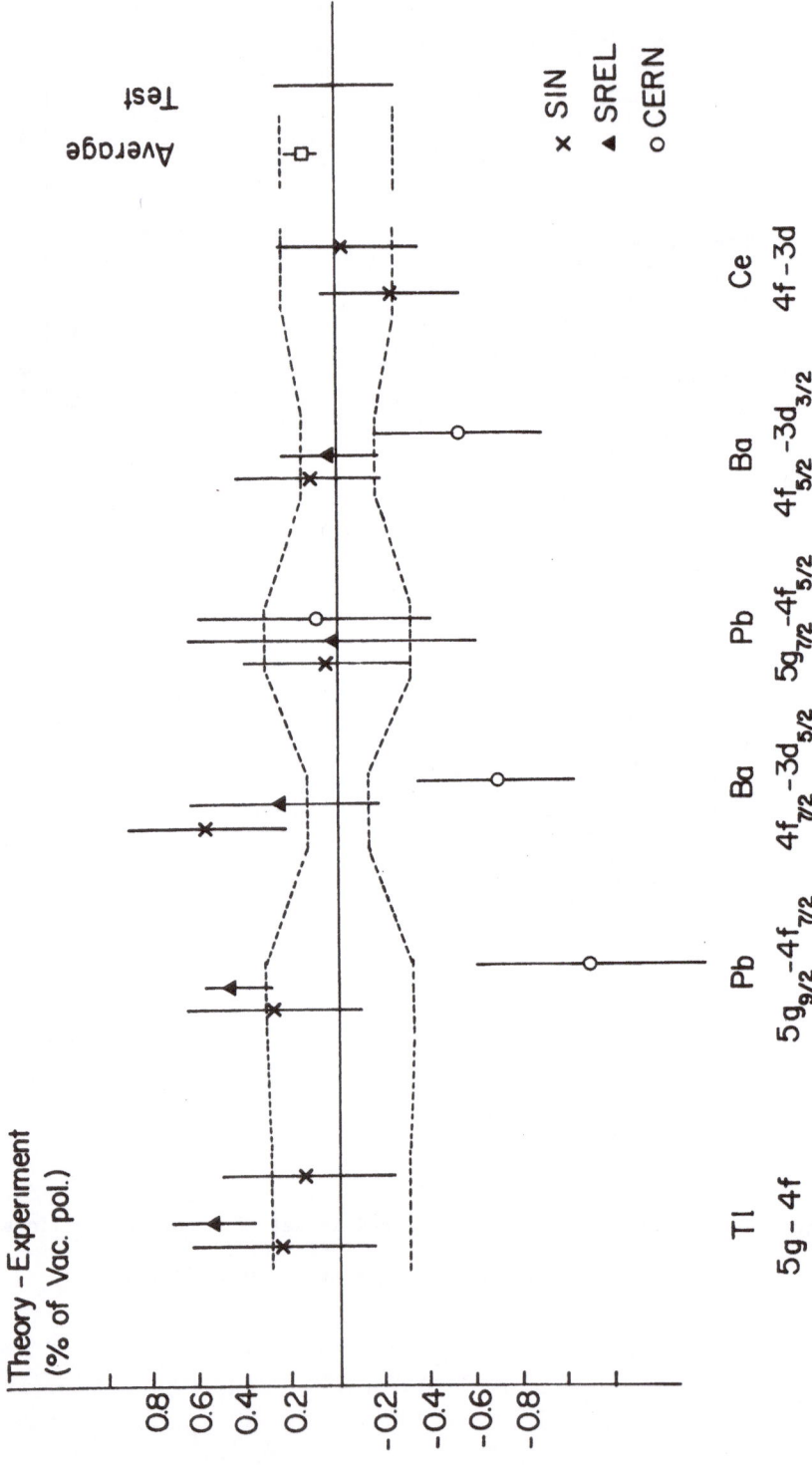

Fig. 4 : Deviations as defined for Fig. 2.

Experimentally several improvements can be envisaged.

Crystal spectrometer:

- better statistics [factor of $\sim$ 3(?)];

- better analysis by eliminating the line shape problem [factor of 2(?)];

- higher orders (??).

This could lead to a total experimental accuracy of 3-4 ppm or a test quality of 0.05-0.1% of the vacuum polarization.

Ge-diode spectrometers:

- elimination of background problems by better resolution;

- improvement of peak-to-background ratio by an anti-Compton device as used by Beetz et al.[13];

- reduction of background lines by coincidence measurements of different cascade transitions;

- reduction of calibration systematics by improving the calibration methods used by the CERN group.

A total experimental accuracy of 2-3 ppm seems, however, to be the limit of this technique, corresponding to a test of QED to 0.04-0.06%. In order to reach a similar precision from the "theoretical" side, significant improvements on the electron screening, nuclear polarization, and also from the QED computations are needed.

In summary I think that the crystal spectrometer technique has a good chance to reach a test quality of 0.05-0.1% for the vacuum polarization; whereas for the heavy muonic atoms, even minor improvements will require enormous efforts.

REFERENCES

1) H.G. Dehmelt, "Anomalie der magnetischen Momente von Elektron und Positron". This conference.

2) J. Bailey et al., Nucl. Phys. B150, 1 (1979).

3) S.R. Lundeen and F.M. Pipkin, Phys. Rev. Lett. 34, 1368 (1975).

4) G. Carboni et al., Nucl. Phys. A278, 381 (1977).

5) E. Zavattini, Proc. 7th Int. Conf. on High-Energy Physics and Nuclear Structure, Zurich, 1977 (ed. M. Locher) (Birkhäuser-Verlag, Basle, 1977), p. 49.

6) H.J. Leisi, Proc. 8th Int. Conf. on High-Energy Physics and Nuclear Structure, Vancouver, 1979 (eds. D.F. Measday and A.W. Thomas), Nucl. Phys. A335, 3 (1980).

7) T. Dubler et al., Nucl. Phys. A294, 397 (1978).

8) L. Tauscher et al., Z. Phys. A285, 139 (1978).

9) H.J. Leisi, Proc. 1st course of the Int. School of Physics of Exotic Atoms, Erice, 1977 (eds. G. Fiorentini and G. Torelli) (Servizio Documentazione dei Laboratori Nationali di Frascati, Frascati, 1977), p. 75.

10) E.G. Kessler et al., Nucl. Instrum. Methods 160, 435 (1979).

11) C.K. Hargrove et al., Phys. Rev. Lett. 39, 307 (1977).

12) E.G. Kessler et al., Phys. Rev. Lett. 40, 171 (1978).

13) R. Beetz et al., Z. Phys. A286, 215 (1978).

THE DETERMINATION OF LAMB SHIFT FROM THE ANISOTROPY
QUENCHING RADIATION FROM METASTABLE
HYDROGENIC ATOMS

P.S. Farago[*]

Department of Physics

University of Edinburgh

Edinburgh, Scotland

Abstract

The measurement of the anisotropy in the quenching radiation emitted by metastable hydrogenic systems in an electric field can be used to determine the Lamb shift. This approach was conceived as a method for the measurement of Lamb shift in high-Z ions but was tested on H, D and He$^+$ where high precision Lamb shift values are available both from theory and from earlier experiments.

After a brief outline of the principle of the method, the main experimental problems are surveyed and the results are summarized.

Introduction

Hydrogenic systems in the metastable 2s state can be induced to radiate by the application of a static electric field. Fite and co-workers (1968) and Casalese and Gerjuoy (1969) first pointed out that this "quenching radiation" is polarized but it went unnoticed at that time that the radiation intensity summed over two orthogonal polarization states is anisotropic. This feature of the radiation pattern emerged from the investigations of Drake and Grimley (1973). They considered the static electric field as a beam of very low frequency polarized in the direction of the field and showed that the transition rate for the emission of quenching radiation of linear polarization $\underline{e}$ in the presence of a static electric field of direction $\underline{E}$ is

---

[*] Currently on leave of absence at the FOM-Institute for Atomic and Molecular Physics, Amsterdam, The Netherlands.

$$|Q|^2 \propto |\underline{E} \cdot \underline{e}|^2 |A|^2 + |\underline{E} \times \underline{e}|^2 |A'|^2 \tag{1}$$

where A and A' are time dependent quantities determined by the fine and hyperfine structure of the atomic system under consideration and are functions of the strength of the quenching field. It should be stressed that $A \neq A'$ only if the Lamb shift is different from zero. Considering observable phenomena, Eq. (1) has two interesting implications.

(1)   If the quenching radiation is observed at right angles to the electric field, the intensity of the radiation polarized linearly at an angle Φ relative to the direction of the electric field will be:

$$I(\Phi) \propto |A|^2 \cos^2\Phi + |A'|^2 \sin^2\Phi$$

Hence a comparison of the intensities polarized parallel and at right angles to the direction of the quenching field gives a "polarization":

$$P \equiv \frac{I_{\parallel} - I_{\perp}}{I_{\parallel} + I_{\perp}} = \frac{|A|^2 - |A'|^2}{|A|^2 + |A'|^2} \tag{2}$$

where the subscripts $\parallel$ and $\perp$ correspond to $\Phi = 0$ and $\Phi = \pi/2$ respectively.

(2)   The total quenching radiation emitted at an angle θ relative to the direction of the quenching field, i.e. summed over two orthogonal states of polarization, will have an intensity:

$$I(\theta) \propto (|A|^2 + |A'|^2)\sin^2\theta + 2|A'|^2\cos^2\theta \tag{3}$$

In other words, the radiation is anisotropic and a comparison of the intensities emitted parallel and at right angles to the quenching field gives an "asymmetry":

$$R \equiv \frac{I(0) - I(\pi/2)}{I(0) + I(\pi/2)} = \frac{|A'|^2 - |A|^2}{3|A'|^2 + |A|^2} \tag{4}$$

The polarization P was measured by Ott et al. (1970) for hydrogen. In the present review we shall be concerned with problems relating to the measurement of the asymmetry R.

It should be stressed that both phenomena, the polarization and the anisotropy of the quenching radiation, hinge on the very existence of a finite Lamb shift because $A = A'$ if the Lamb shift vanishes. Both quantities, P and R, are approximately proportional to the Lamb shift and, in principle, they yield identical information since

P = 2R/(R-1). Yet anisotropy measurements are preferable in practice because this quantity can be determined more accurately than polarization at the vacuum ultraviolet wavelength of the quenching radiation.

From the outset it had been expected that it would be possible to measure R and hence to determine the Lamb shift to an accuracy of at least 0.1%. Such an accuracy falls short of that obtained by microwave resonance techniques applied to H, D or He$^+$; but it makes the anisotropy method a serious candidate for measurements on heavy ($Z > 3$) ions because, if realized, such an accuracy would be quite adequate to distinguish between competing theoretical predictions. Compared to the alternative approach of quenching rate measurements the anisotropy method has some inherent advantages. For example the intensity of the quenching field need not be known as accurately because the effect is independent of field strength in the limit of weak fields; there is no need for tracking the intensity of the radiation over several decay lengths with the consequent loss of intensity and uncertainties introduced by beam deflection. Even where tunable lasers are available for resonance experiments the large width of the resonance makes the superiority of that approach arguable.

For the derivation of an empirical value of the Lamb shift from the measured asymmetry Drake and Grimley (1975) developed a non-pertubative theory of the quenching process. This theory allows the calculation of a "theoretical" asymmetry R_o using the theoretical value of the Lamb shift and accepted spectroscopic data as input parameters. A comparison between the calculated asymmetry, R_o , and the measured asymmetry R, yields the experimentally determined Lamb shift (Drake and Lin, 1976). In order to explore the potentials of the anisotropy method and to test the validity of the theory underlying the interpretation of the results, a series of experiments were performed on light species, namely H, D and He$^+$. In each of these cases the Lamb shift is known to high accuracy and the choice of these different species permitted the investigation of different aspects of the phenomena involved.

THE EXPERIMENTAL SCHEME

The experimental apparatus employed throughout the experiments was basically the same (fig. 1). In the hydrogen (deuterium) experiments a mono-energetic (typically 10 KeV) beam of protons (deuterons) traverses a cesium vapor cell; the emerging beam contains metastable hydrogen (deuterium) atoms in the 2s state produced in a near-resonant charge exchange reaction. The remaining ions are deflected out of the beam by

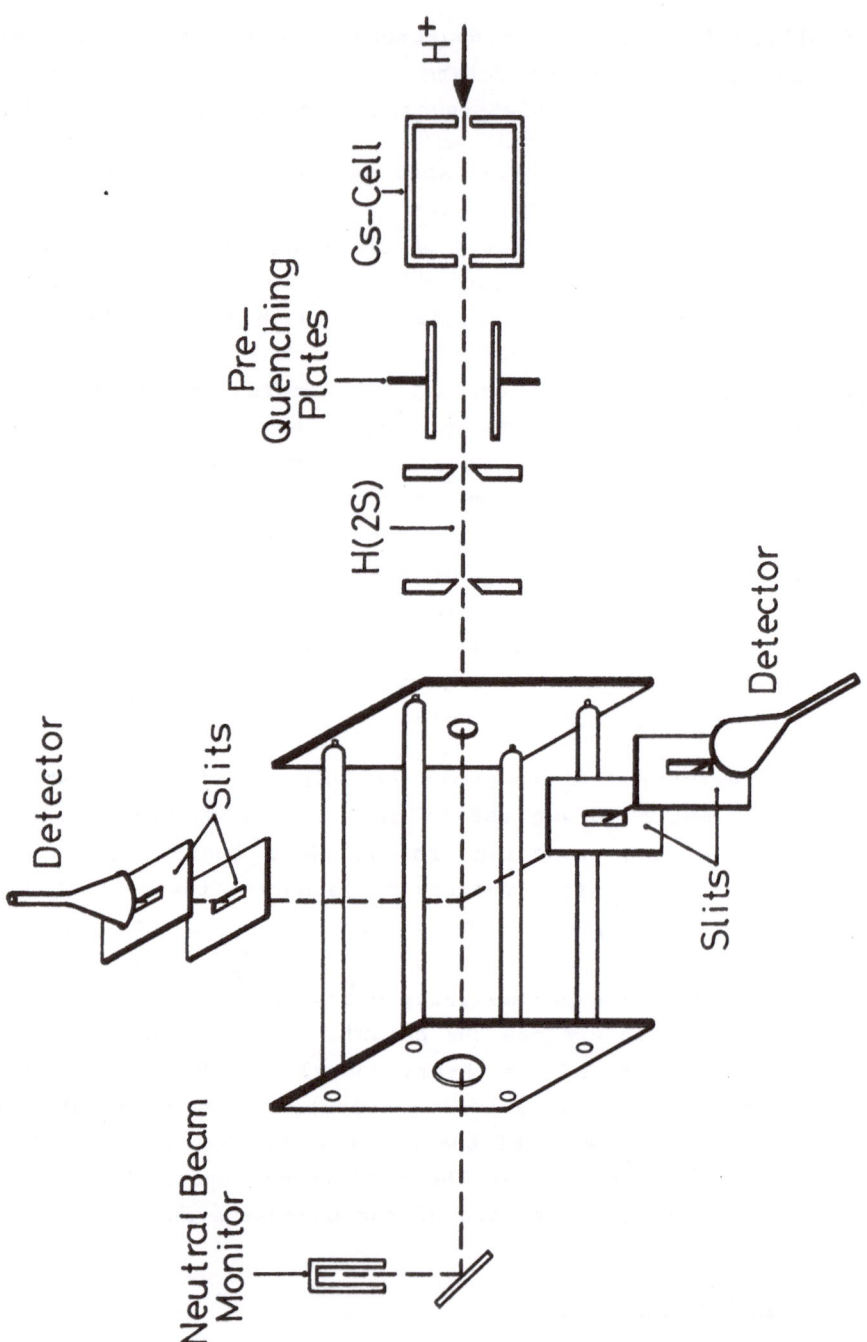

Fig. 1

tne weak (∿ 10 V/cm) electric field between the prequenching plates. A collimated beam of neutrals then enters the observation region, where the metastable atoms are quenched in an essentially uniform electro-static field. In the He$^+$ experiments the metastables are produced by passing 90 keV ground state ions through a gas cell. Prequenching, when required, is induced by a longitudinal field. After collimation the ion beam enters the quenching field.

The quenching field is maintained by four cylindrical rods, as in a quadrupole lens, arranged symmetrically with respect to the beam axis. In this case, however, adjacent rods form a pair kept at the same potential. To enhance the symmetry of the system it is surrounded by a cylindrical mantle (not shown in the sketch) centered on the symmetry axis and held at the same potential as the end plates.

A short section of the beam in the central region of the quenching field is viewed by two channeltrons detecting Ly-α radiation emitted in mutually perpendicular directions; for a given choice of electrode potentials one is parallel and the other is perpendicular to the direction of the quenching field. Equation (3) shows that the total intensity I(θ) is very insensitive to small errors at angular settings in the neighbourhood of θ=0 and π/2 , making it relatively easy to correct for the effect of a finite solid angle of observation.

There are two main sources of instrumental asymmetry: (i) the two detectors have somewhat different acceptance angles and detection efficiencies and (ii) the presence of stray magnetic fields, $\underline{B}$, gives rise to a motional electric field $\underline{E} = \underline{v} \times \underline{B}$. ($\underline{v}$: particle velocity).

Both these effects can be eliminated to a high degree of accuracy by rotating the quenching field in steps of π/2 relative to the di-rections defined by the line-of-sight of the fixed detectors. This is achieved by a cyclic change of the polarity of the quadrupole rods and by recording the counting rates measured simultaneously by the two detectors at each of the four consecutive field orientations: $N_1(\Phi)$ and $N_2(\Phi+\pi/2)$ with $\Phi=0$, π/2 , π , 3π/2 .

The effect of the motional field could be detected by reversing the direction of the quenching field, ẏielding a small discrepancy:

$$N_1(0)/N_2(\pi/2) \neq N_1(\pi)/(N_2(3\pi/2)$$

In order to eliminate this effect the measured quenching radiation intensity is defined as the mean of two values obtained at field directions reversed. Thus:

$$N_1' = 1/2(N_1(0) + N_1(\pi)) = a_1N(0)$$

$$N_2' = 1/2(N_2(\pi/2) + N_2(3\pi/2)) = a_2N(\pi/2)$$

and, similarly:

$$N_1'' = 1/2(N(\pi/2) + N_1(3\pi/2)) = a_1N(/2)$$

$$N_2'' = 1/2(N_2(\pi) + N_2(0)) = a_2N(0)$$

where N(0) and $N_2(\pi/2)$ are the apparent rates of emission parallel and at right angles to the quenching field and a_1 and a_2 are constants determined by the angle of acceptance and efficiency of the two detectors respectively. From the above equations the asymmetry analogous to that defined by Eq. (4) is obtained in the form:

$$R' = (r-1)/(r+1) \tag{5}$$

where

$$r^2 = N_1'N_2'' / N_2'N_1''$$

The effect of intensity fluctuations on the results which involve pairs of counting rates measured at different times were minimized by monitoring the neutral beam current with the aid of a current-to-frequency converter to define the counting period for each measurement.

The measured counting rates $N_j(\Phi)$ contain a contribution from background noise. In order to take this into account a high (1500 V/cm) prequenching field was applied and the isotropic component of the still observable radiation was determined. The noise thus defined amounted to about 1% of the signal obtained in the absence of the prequenching field and gave a small correction to the directly measured asymmetry R'.

It should be mentioned that in the H and D experiments the noise could readily be kept at this low level but in the He$^+$experiments special precautions were required.

The main source of experimental error arises from the random fluctuations in counting statistics. If the ratio of the sensitivities of the two detector systems is denoted by a, and n denotes the total number of counts, the standard deviation in the observed asymmetry is:

$$\sigma_R = \frac{1}{2} (a^{1/2} + a^{-1/2}) \{(1-R'^2)/n\}^{1/2}$$

The uncertainty in the noise measurements makes a small additional contribution to the statistical error. If the signal and noise measurements alternate and the signal-to-noise ratio is s, the total standard deviation becomes:

$$\sigma = \sigma_R \{1 + s^{-1}(1+R'^2)(1-R'^2)^{-1}\}^{1/2} \tag{6}$$

The directly observed asymmetry R' requires some small corrections due to systematic effects and the uncertainty in these corrections must also be assessed.

(1) The field strength in the neighbourhood of the beam must be calculated. In principle this is a simple task, but the required accuracy (better than 0.1%) leads to non-trivial computational problems as discussed in detail by van Wijngaarden and Drake (1978). The precision to which the field is known depends, in the end, on the tolerances to which the apparatus is fabricated, and some uncertainty introduced by fringing field effects.

(2) The finite acceptance angle of the detectors is taken into account by integrating the radiation intensity as a function of direction over the source and over the solid angle determined by slits between the radiating beam of metastables and the detectors.

(3) A loss of counts arises from electronic dead-time in the detection.

(4) In experiments with ion beams their deflection in the quenching field is accompanied by a small change in the apparent anisotropy. This arises partly from the change of position of the effective radiation source relative to the detectors and partly from a change in the direction of observation. These two factors work in opposition but do not cancel one another.

EXPERIMENTAL RESULTS

(a) Hydrogen (Drake, Farago and van Wijngaarden, 1975).
The quenching of the metastable $2s_{1/2}$ state of hydrogen is a rather complicated phenomenon because of the rather strong hyperfine coupling. If a H-atom in one of its four possible $2s_{1/2}$ hyperfine states enters an electric field it induces a mixing with twelve possible intermediate $2p_{1/2}$ and $3p_{3/2}$ hyperfine states. From any of these states the atom can return to its ground state by emitting a photon. The atoms entering the quenching field "sees" the perturbation turned on in a finite period of

time. If this period is short relative to the response time of the atom the intensity of the radiation emitted into a small solid angle shows "quantum beats". Measured as a function of the elapsed time after entering the field the asymmetry R' too shows large oscillations near the Lamb shift frequency modulated at the hyperfine frequency of the $2s_{1/2}$ state, and each peak is further structured by rapid oscillations near the fine structure frequency. The oscillations decay in 20 ns or so, and beyond this period the anisotropy shows only weak time dependence arising from the different rates at which the different hyperfine components of the perturbed metastable states decay.

In the experiments the field was "switched on" over a period of a few nanoseconds which is fast compared to the life time of the $2s_{1/2}$ state but short compared to the life time of the 2p states. Measurements were performed in the regime of slowly varying anisotropy. In order to avoid the need of precise field calibration the anisotropy R (corrected for finite solid angle and noise) was determined at a sequence of different quenching field strengths E, and a curve:

$$R = \sum_{k=0}^{n} a_k E^{2k}$$

was fitted to the experimental results. In performing the least-squares fitting to single runs the approximation to n=2 was always found significant. While curves obtained for individual runs were slightly different in details, they always led to extrapolated values R(E=0) which agreed with one another within their margins of error.

In the limit of zero field strength the fractional error in the asymmetry was 8.6×10^{-4}; a summary of the numerical results is contained in Table 1.

The effect of finite transit time is demonstrated in fig. 2. The two experimental points were obtained at the same quenching field with H-beams of different kinetic energies; the curve represents the theoretical prediction. A more rigorous test of the theory used in the interpretation of the experimental results was carried out by a careful study of the quantum beats mentioned earlier (van Wijngaarden et al., 1976).

For these experiments a special technique was developed by means of which the perturbation can be switched on "suddenly". It is an essential feature of this technique that the use of a beam-foil is avoided and this makes it safe to assume that the initial state of the

Table 1: Summary of results.

Species	ANISOTROPY METHOD				To be compared with	
					Other expt.	Theory
	Quenching field (V/cm)	Time of flight (ns)	Asymmetry R	Lamb shift (MHz)	Lamb shift (MHz)	Lamb shift (MHz)
H(2s)	→0	25	0.13901 ± 0.00012	1057.4 ± 1.0	1057.90 (a) ± 0.06	1057.91 (d) ± 0.01
	151.6	19.9	0.15093 ± 0.0002	1057.51 ± 1.6	1057.86 (b) ± 0.06	
	151.6	132.4	0.15172 ± 0.0002	1057.4 ± 1.6		
			Avarage of above	1057.03 ± 0.9		
D(2s)	81.613	68.33	0.144259 ± 0.000021	1059.36 ± 0.16	1059.24 (c) ± 0.06	1059.271 (d) ± 0.025
					1059.00 (b) ± 0.06	1059.241 (e) ± 0.027
He$^+$(2s)	599.4		0.118150 ± 0.000024	14040.2 ± 2.9	14040.2 (f) ± 1.8	14042.05 (e) ± 0.55
					14046.2 (g) ± 1.2	14044.78 (d) ± 0.61

References: (a) Robiscoe and Shyn, 1970   (d) Erickson, 1971   (g) Narasimham and Strombotne, 1971
(b) Triebwasser et al.1953   (e) Mohr, 1976
(c) Cosens, 1968   (f) Lipworth and Novick, 1957

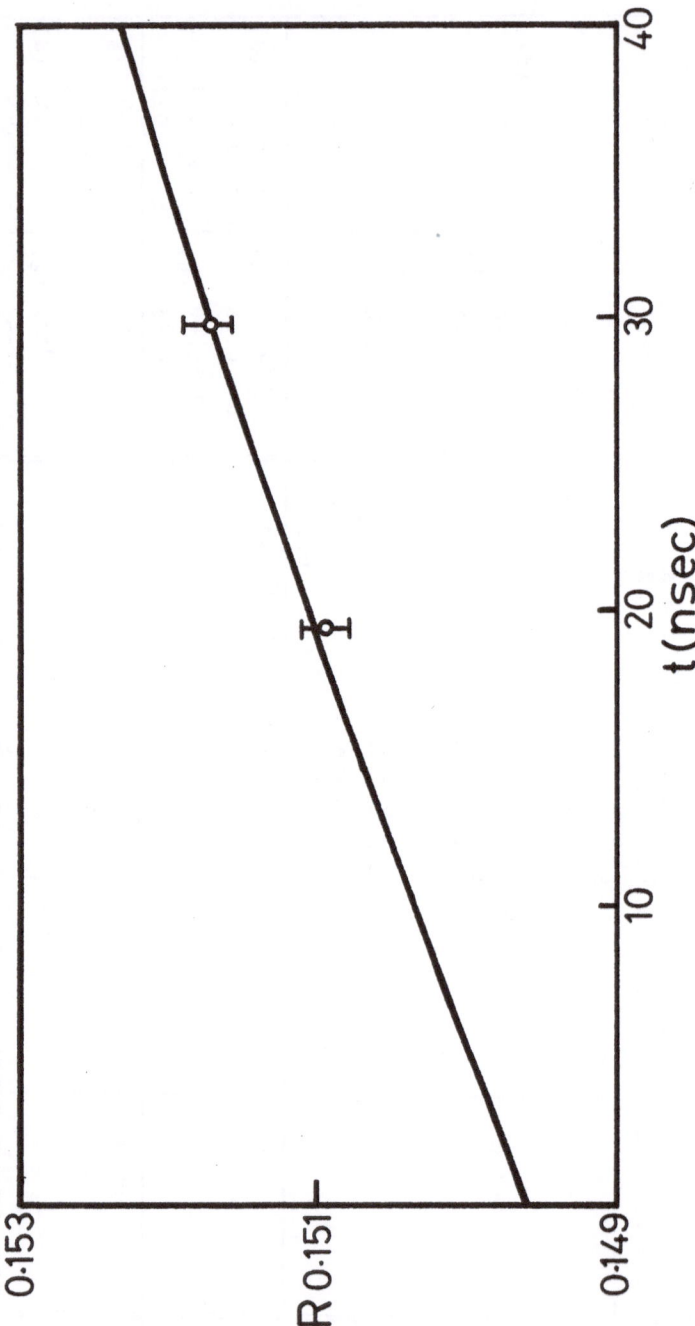

Fig. 2

atom is an incoherent mixture of all four $2s_{1/2}$ hyperfine states with equal statistical weights. In contrast, the initial state amplitudes in a beam emerging from a beam foil are not adequately known.

The experimental results agreed well with calculations assuming sudden excitation. Various frequency components of the time dependent radiation intensity were identified with specific hyperfine transitions or groups of transitions.

(b) Deuterium (van Wijngaarden and Drake, 1978).
In an endevour to improve the accuracy of the anisotropy measurement the choice of deuterium was preferable to that of hydrogen because in the former the hyperfine coupling is much weaker and hence the time dependent effects are smaller.

Since the dominant contributions to the experimental error arises from counting statistics, the most efficient use of the measuring time is of utmost importance. For this reason asymmetry measurements were performed at a single quenching field strength. The experimental para-meters were chosen in such a manner that in the observation region, about half way down the axis of the quenching field, the time depend-ence of the asymmetry was very small (as a function of the position x of the effective source of radiation the asymmetry varies at a rate $\delta R/\delta x \sim 2 \times 10^{-5}$ cm^{-1}) and the intensity of the radiation was approxi-mately a maximum at the chosen field strength. Thus errors due to small field intensity variations and to small spread in the kinetic energy of the particles were minimized.

In order to obtain high absolute precision in the empirical Lamb shift value derived from measurements at finite quenching field strength, the accuracy of instrumental parameters must be consistent with the assumptions made in the calculations.

In this experiment the critical dimensions of the quadrupole system which maintains the quenching field, and of the slit system which selects the observed radiation were maintained to a tolerance of 5 μm.

Contributions from different sources of experimental error are listed in Table 2. The validity of the estimate of the dominant random error was tested by a statistical analysis of the 2038 individual asymmetry measurements, each containing, on the average, 1.355×10^{6} counts. The fractional error in the empirical Lamb shift value (see Table 1) is 2.7-times larger than that of the best results obtained by

Table 2: Sources of error in the anisotropy measurement due
to uncertaintie in various parameters

Parameter	$\delta R/R$ (ppm)	
	D(2s)	He$^+$(2s)
-Counting statistics	138	204
-Electronic dead-time correction	26	10
-Solid angle correction	5	5
-Electric field strength	24	3
-Fringing field effects	8	-
-Beam deflection correction	-	3
$(\Sigma_i \delta R_i^2)^{1/2}$ /R	143	204

microwave resonance techniques.

(c) Helium ions (Drake, Goldman and van Wijngaarden, 1979).
Since the anisotropy measurements were conceived as an approach to
determine the Lamb shift in high-Z hydrogenic ions, it was important to
explore those problems which arise in working with charged as opposed
to neutral beams. For this purpose the choice of He$^+$ is ideal, not only
because the Lamb shift is known to a high accuracy but also because the
theoretical calculation of the anisotropy is made relatively simple by
the absence of hyperfine structure in ^{4}He. In the absence of a magnetic
field there is no Zeeman splitting either, so the anisotropy is
independent of time apart from the rapidly decaying transient effects
associated with the onset of the perturbation at entry to the quenching
field.

This experiment quickly revealed a major difficulty inherent in
experiments with fast ion beams, namely the problem of keeping the
noise at an acceptable level. The fast ion beam as it passes through
the residual gas in the quenching cell produces an abundance of charged
particles, including metastable ions, which are accelerated by the
quenching field. If they were allowed to enter the detectors (in this
experiment channeltrons with open cones) not only excessive noise would
be produced but its magnitude would depend on the direction of the
quenching field yielding a false asymmetry. Therefore a filter system
was inserted in the line of sight of detectors which rejects the
charged particles at the cost of a moderate loss in photon counts. The
filter consists of a pair of electrostatic deflector plates bracketed
by two collimating slits and followed by two more apertures, the first
of which is covered by a thin Formvar film. With the aid of this
arrangement a signal-to-noise ratio of about 100:1 could be achieved.

The fractional error in the asymmetry measurement is 204×10^{-6} ,
made up of contributions as listed in Table 2. The empirical Lamb shift
value derived from the measured asymmetry is accurate to about the same
margin of error. A comparison of theoretical and experimental values is
given in Table 1. It should be noted that the error margin of this
measurement is comparable with those obtained in the most precise
measurements performed to-date. The actual value derived from the
asymmetry measurement agrees with the results of Lipworth and Novick
(1957), and the calculations of Mohr (1975), but there is a distinct
discrepancy in comparison with the experimental result of Narasimham
and Strombotne (1971) and the calculations of Erickson (1971).

CONCLUDING REMARKS

Experimental tests on H-, D- and He$^+$-beams in the metastable 2s state show that the measurement of the anisotropy of the quenching radiation as a means of determining Lamb shift stands up to expectations. It was stressed that this method was first conveived as a suitable approach to Lamb shift measurements in high-Z hydrogenic ions. Although there are at least two laboratories where such an application of this method is currently in progress, there are as yet no results available for critical assesment. Judging by the example of the He$^+$-experiments it is clear that the noise problem is more difficult when high energy high-Z hydrogenic ions are involved. It is more difficult to produce "clean" beams of the required species and the required quenching field is much higher than those applied in the experiments described. Both these factors aggravate the task of reducing the noise level at residual gas pressures of conventional high vacua. Difficulties arising from the high intensity of the quenching field may be circumvented by applying a magnetic field $\underline{B}$ such that the "motional field" $\underline{v} \times \underline{B}$ causes the required perturbation of the metastable state. No doubt this approach is not free of difficulties either. Yet it seems reasonable to expect that, where anisotropy measurements are feasible, the margin of error will stay low enough to make the results significant.

ACKNOLEDGEMENT

The work reviewed above was supported by the National Research Council of Canada and, in part, by a NATO travel grant. The author is also grateful for a Canadian Commonwealth Research Fellowship he held and for the hospitality he enjoyed in the Physics Department at the University of Windsor, Windsor, Ontario, Canada, making it possible for him to take part in the early stages of the development of this project.

REFERENCES

Casalese J.S. and Gerjuoy E. 1969 Phys. Rev. <u>180</u>, 327.

Cosens B.J. 1968 Phys. Rev. <u>173</u>, 49.

Drake G.W.F. and Grimley R.B. 1973 Phys. Rev. <u>A8</u>, 157.

Drake G.W.F. and Grimley R.B. 1975 Phys. Rev. <u>A11</u>, 1614.

Drake G.W.F., Farago P.S. and van Wijngaarden A. 1975,
    Phys. Rev. <u>A11</u>, 1621.

Drake G.W.F. and Lin C.P. 1976, <u>A14</u>, 1296.

Drake G.W.F., Goldman S.P. and van Wijngaarden A. 1979,
    Phys. Rev. <u>A20</u>, 1299.

Erickson G.W. 1971, Phys. Rev. Lett. <u>27</u>, 780 (see also: J. Chem. Phys.
    Ref. Data <u>6</u>, 831, 1977).

Fite W.L., Kauppila W.E. and Ott W.R. 1968, Phys. Rev. Lett. <u>20</u>, 409.

Lipworth E. and Novick R. 1957, Phys. Rev. <u>108</u>, 1434.

Mohr P.J. 1976, in <u>Beam Foil Spectroscopy</u>, ed. I.A. Sellin and
    D.J. Pegg (Plenum, New York, 1976) p. 89.

Narasimham M. and Strombotne R. 1971, Phys. Rev. <u>A4</u>, 14.

Ott W.R., Kauppila W.E. and Fite W.L. 1970, Phys, Rev. <u>A1</u>, 1089.

Robiscoe R. and Shyn T. 1970, Phys. Rev. Lett. <u>24</u>. 559.

Triebwasser S., Dayhoff E.S. and Lamb Jr. W.E. 1953, Phys. Rev. <u>89</u>, 77.

Van Wijngaarden A., Goh E., Drake G.W.F. and Farago P.S. 1976,
    J.Phys. B. (Atom. Molec. Phys.) <u>9</u>, 2017.

Van Wijngaarden A. and Drake G.W.F., 1978 Phys. Rev. <u>A17</u>, 1366.

THE LAMB SHIFT OF THE HYDROGEN ATOM AND HYDROGENIC IONS

R. Wallenstein
Fakultät für Physik
Universität Bielefeld
48 Bielefeld, West Germany

I Introduction

For the development of quantum mechanics and quantum electrodynamics (QED), studies of atomic hydrogen have played an important role. Since only a single electron and proton coupled by the radiation field have to be considered the dynamics of this simple atomic system can be calculated with high precision and may be compared directly with experimental measurements. The difficulties in reconciling the theoretical predictions with experimental results have been an effective stimulus to the development and improvement of both the theoretical models and the experimental investigations.

A good example for such interaction between theory and experiment are the investigations of the fine structure intervals in atomic hydrogen and deuterium. The quantum mechanical Dirac theory which includes spin and relativistic corrections predicts a degeneracy for states with same quantum numbers n and j. In their famous experiment Lamb and Retherford[1] proved, however, that the $2 S_{1/2}$ and $2 P_{1/2}$ state are in fact not degenerate. Subsequently the splitting between these states, the so-called "Lambshift" played a major part in the development of a new theory, called quantum electrodynamics.

The principles and the results of the QED and the theory of the Lamb shift[2,3] are discussed in several contributions in this volume. For the discussion of experimental results it will, therefore, suffice to mention only a few general points.

The Lamb shift is made up of several contributions which are the result of (a) zero point oscillations of the quantized electromagnetic field (self energy); (b) vacuum polarization; (c) the anomalous magnetic moment of the electron (which acts through the LS-coupling) and (d) nuclear size and nuclear structure effects. For contributions of the kind (a) and (b) the electron wavefunction has to be nonzero at the origin (nucleus). Essentially this is the case only for S states which therefore experience by far the largest shift. To this shift the dominant contribution comes from (a). The displacements of states with L > 0 are considerably smaller and mainly caused by (c) with some addition from (a) and (b). Contributions from (d) are always small although not negligible.

The calculation of the Lamb shift is performed by evaluating the various contributions in orders of α, $(Z\alpha)$, and m/M.[4,5] The largest contribution to the Lamb shift of S states comes from the one-photon self-energy term which has been calculated explicitely in orders of $(Z\alpha)$ up to terms of $(Z\alpha)^6$. For large values of $Z (\geq 10)$ the series expansion in $(Z\alpha)$ does not converge very well. Recently Erickson[6] and Mohr[7] independently calculated the expansion for higher Z without truncation. In turn this allowed the estimation of the uncalculated high-order terms for low Z. Although the results of the two calculations[6,7] are somewhat different, the discrepancy is not sufficiently large so that the best present experimental results could decide between them outside of their experimental uncertainties.

The Z dependence of the $S_{1/2} - P_{1/2}$ Lamb shift interval is somewhat slower than the Z^4 dependence of the fine structure intervals calculated by the Dirac theory. This is caused mainly by the leading term which is proportional to $\alpha(Z\alpha)^4\{\ln[1/(Z\alpha)^2]-C\}$ with $C \approx 1/4$ of the ln term for Z = 1. The dependence given by this term is slightly modified by contributions of higher order in $(Z\alpha)$, which become rapidly more important with increasing Z.

For example, the term with $\alpha(Z\alpha)^5$ contributes 0.7 % of the total Lamb shift $2 S_{1/2} - 2 P_{1/2}$ in H(Z = 1) compared with 7.1 % in hydrogenic carbon $^{12}C^{5+}$ (Z = 6). The corresponding contributions from $\alpha(Z\alpha)^6$ are 0.04 % for H and 1 % for $^{12}C^{5+}$. Therefore measurements for the ion $^{12}C^{5+}$ with an accuracy of approxomately 1 % are already very useful for a test of the theory. Therefore it appears to be very important to extend the measurements to as high values of Z as possible.

For S states the n dependence of the Lamb shift follows very closely the $1/n^3$ dependence of the Dirac fine structure. The small Lamb shifts with j > 1/2 also follow closely both the Z^4 and $1/n^3$ dependence of the Dirac splittings. For such states however there are only a few, not particularly accurate experimental results available.

The ultimate experimental accuracy which can be obtained strongly depends on the width of the observed signals. This is because even a well-known signal shape allows the signal position to be determined only to a certain fraction of the width. For highest possible precision Z- and n-values should be selected which provide the smallest signal width in comparison with the interval to be investigated. The smallest experimental signal width which can be obtained for a particular state is limited of course by the natural widths of the states involved. Unfortunately the Z and n dependences of the natural width are almost exactly counterbalanced by the Z and n dependences of the corresponding Lamb shift intervals so that not much can be gained in this way for states with different n.

Over the last 30 years numerous experimental investigations of the Lamb shift have been carried out with a variety of different methods. A rather complete list and detailed descriptions of various experimental investigations can be found in the excellent review of ref. 8. Many of these measurements were triggered by the somewhat unsatisfactory basis of the theory and by discrepances between theory and experiments or between various experiments. Today a very high level of accuracy has been reached and with a few exceptions, there is satisfactory agreement among the experiments and with theory.

II   Precision measurements of the Lamb shift $2\ ^2P_{1/2} - 2\ ^2S_{1/2}$ in H and D.

Since the signal to noise ratio is best for low Z and n and because low n states are less sensitive to external perturbations the most accurate measurements have been performed for the $2\ P_{1/2} - 2\ S_{1/2}$ shift in H and D. The most precise results were obtained in resonance experiments using radio frequency, anticrossing and level crossing methods in slow beam investigations.

The basic experimental set up for such measurements - as it was first used by Lamb and coworkers[9-14] - is shown in fig. 1. A beam of H and D atoms is created by thermal dissociation of H_2 or D_2 in a hot tungsten nozzle. A crossed electron beam excites some of the atoms. Except the S states all other states decay before the beam enters the rf region inside of an electromagnet. In the static magnetic field transitions are induced by the rf-field from the $2 S_{1/2}$ to the rapidly decaying $2 P_{1/2}$ when the energy separation between suitable Zeeman sublevels of S and P is tuned by the external magnetic field to match the fixed frequency of the rf-field. The surviving 2 S atoms of the beam are selectively detected by a detector sensitive to metastable 2 S atoms.

In the experiments of Lamb et al. the transitions αe and αf (fig. 2) were studied at a field of about 700 Gauss. Since the natural lifetime of the 2 P state of $1.6 \cdot 10^{-9}$ sec corresponds to a level width of 100 MHz a width of the resonance curves of about 120 MHz would be expected for a partly saturated signal. Unfortunately in hydrogen the hyperfine structure (hfs) of the 2 S state is of the same order so that the two hfs components $\alpha^+ f^+$ and $\alpha^- f^-$ of the signal can not be fully resolved and formed a double peak signal.[13] D has a smaller hfs than H so that the composite signal comes close to a single component situation.[13] D atoms also travel more slowly because of their larger mass so that velocity dependent effects on the signal are reduced. Thus the necessary corrections to the final results were considerably smaller for D than for H and a precision measurement of the Lamb shift was carried out only for D.

The most important corrections to the experimental value were due to the overlap of the hfs-resonances, to rf power shifts, to magnetic field inhomogenities, the Zeeman curvature of the substates, the variation of electric dipole matrix element with the magnetic field, to stark shifts of the resonance center due to stray and motional electric fields and due to the velocity distribution and changes of this distribution of the beam in the magnetic field.

In the experiments performed by Lamb and coworkers the accuracy of the final results was ± 0.1 MHz which corresponds to 10^{-3} of the natural linewidth of 100 MHz. The total corrections applied were between - 0.3 MHz (D) and - 3.2 MHz (H).[13]

In a second series of experiments the amount of corrections required for the results was reduced considerably. This was achieved by hyperfine selection of the metastable beam before it enters the

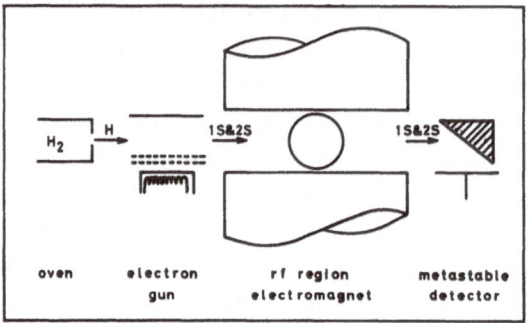

Figure 1   Scheme of the experimental set up for slow beam investigations
           of the 2 $S_{1/2}$ - 2 $P_{1/2}$ splitting in H and D.[1,9]

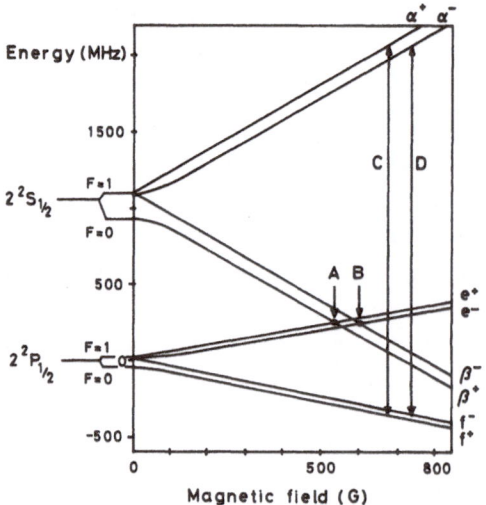

Figure 2   Zeeman effect of the 2 $^2S_{1/2}$ and 2 $^2P_{1/2}$ states of hydrogen
           including hyperfine structure (A_{hfs} = 177. 56 MHz for $S_{1/2}$
           and 55.19 MHz for $P_{1/2}$). The rf-transitions C and D have been
           used by Triebwasser et al. (Ref. 13). The anticrossings A and
           B were investigated by Robiscoe and Cosens (Ref. 16, 19).

principal interaction region. In this way the signals were no longer superpositions of several hfs components. In addition the magnetic field was applied parallel to the beam axis, thus reducing the motional electric field experienced by the atoms.

Instead of inducing rf-transitions between Zeeman sublevels the anticrossing method[15] was employed. In this method which was introduced by Robiscoe[16] for the measurement of the Lamb shift in H, the rf field is replaced by a static electric field ($\perp$H). This electric field mixes the states near degeneracy and causes the levels to repel each other so that the actual crossing is removed. Due to the state mixing the lifetime of the S component is effectively reduced. Thus a resonance signal can be obtained near 575 G where the substates β and e cross (fig. 2).

Due to the hfs the anticrossing βe consist of the two allowed components A at 538 G and B at 605 G. One of these components can be suppressed however, if a state selected beam of β^+ and β^- is produced. This is achieved by first applying in the region of electron-excitation a magnetic field of 575 G (near the crossings A and B) perpendicular to the atomic beam axis so that all β states are quickly quenched by the motional electric field. The α branches remain unquenched. These 2 S atoms fly into a region of the apparatus called flopper where the magnetic field is reduced to zero in such a way that the population of the states is redistributed. In zero magnetic field the branches α^+, α^- and β^- degenerate to the F = 1 state. In this way β^- is created. Therefore the 2 S beam entering the magnetic field in the interaction region consists only of α^+. α^- and β^- so that the anticrossing component B (β^-e^-) is observed without interference from A (fig. 3).

The anticrossing signals are very sensitive however to small static fields, such as motional fields caused by a misalignment of the main magnetic field with respect to the atomic beam axis. This and a measurement of the velocity distribution of the metastables resulted in two revisions[17,18] of the original results.[16,19] But even including these revisions the total corrections applied to the result were only of the order of - 0.3 MHz much less than in the rf measurements.[13] In spite of this improvement the accuracy of the final result ($\pm$ 0.1 MHz) did not exceed the value quoted for the rf-investigation.[13]

It should be mentioned that similar measurements were carried out on deuterium with comparable accuracy.[20,21]

In contrast to the time of the early rf-measurements the progress in microwave techniques made it soon possible to maintain a constant power level in the interaction region even when sweeping the frequency over quite a wide range. This new possibility has led to very precise Lamb shift measurements in which the microwave frequency was varied and no magnetic field was applied.

Using this direct rf-method Andrews and Newton[22] carried out one of the most accurate measurements of the $2 S_{1/2} - 2 P_{1/2}$ Lamb shift of H. In their experiment an H beam with an energy of 21 KeV was used. This energy is low enough for the creation of metastables by charge exchange in H_2 gas and to experience an adiabatic switch-on and switch-off of the rf-field, when passing through the microwave region. State selection was achieved by inducing in a strong rf-field $(S_{1/2}, F = 1) \leftrightarrow (P_{1/2}, F = 1,0)$ transitions, which completely quenched the $F = 1$ state of the $2 S_{1/2}$. Subsequently the transition $(S_{1/2}, F = 0) \leftrightarrow (P_{1/2}, F = 1)$ was induced in a second rf-field. The number of metastables surviving in this second rf-region are recorded by Stark quenching and detection of the emitted Lyman-α radiation.

To eliminate Doppler shifts the measurements were taken for both directions of the rf-field and individually corrected for Bloch-Siegert shift[23,24] by about 90 ppm (90 KHz). Additional corrections amounted to + 27 ppm (29 KHz). The overall uncertainty is made up to the larger part by systematic uncertainties and is given as ± 19 ppm (20 KHz), corresponding to 1/5000 of the natural width of the P state.

All resonance experiments described so far were limited in their resolution by the natural widths of the S and P states. Further improvements are extremely difficult since this requires the determination of the signal position to 10^{-4} or even less of the signal width.

In order to obtain a narrowing of the observed magnetic resonance signals Lundeen and Pipkin[25] used the well-known Ramsay-arrangement of two separated rf-fields. For the Ramsay method the main rf-field is split into two sections rf 1 and rf 2 oscillating coherently and separated by a variable gap. The atoms may spend the time τ in each section and the time T in the gap. The resulting resonance signal is dominated by S → P transitions which are induced in the single rf field sections. Superimposed on this signal is an interference pattern. The amplitude and the width of this oscillatory part is determined by the

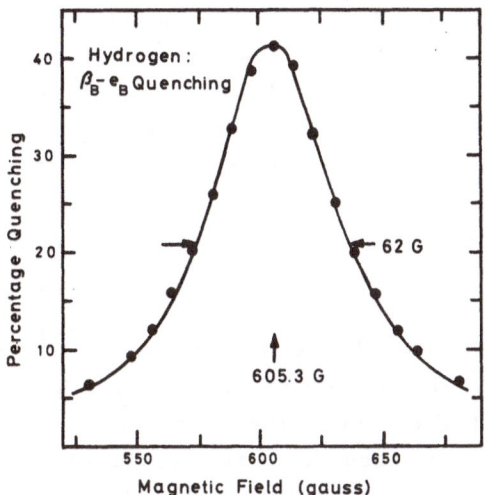

Figure 3 (From Ref. 16) Hyperfine state selected anticrossing signal β⁻e⁻ (marked B in Fig. 2) in hydrogen.

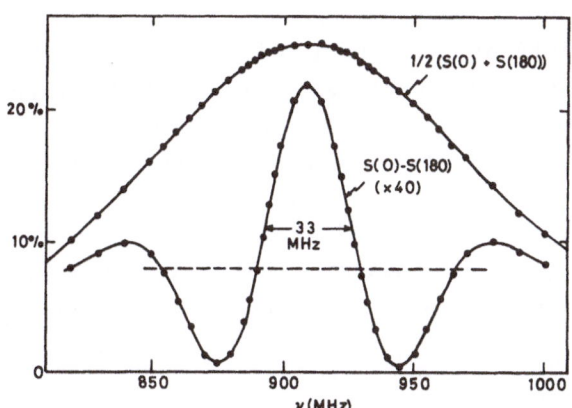

Figure 4 (From Ref. 25) Theoretical and experimental resonance curves of the $2 S_{1/2}$ (F = 0) → $2 P_{1/2}$ (F = 1) transition of H obtained with the separated-rf-field method. (For further details: see text).

time T spent between the rf-field sections. The oscillating part of the signal can be interpreted as an interference term in the intensity of S states detected behind the two rf-field sections which is based on two-amlitudes of S states which spent the time T between the rf-sections as S states or spent this time as P states. With increasing T the number of oscillations increases and their width is reduced. Because the P states rapidly decay the interference term, however, decreases rapidly with increasing T.

In their experiment Lundeen and Pipkin[25] used 2 S atoms of 50 - 100 KeV energy. The number of surviving 2 S atoms was measured as a function of the rf-frequency. The interference could be separated by the total signal S by recording the difference signal between the signal S(0) with the two rf-sections in phase and the signal S(180) with the two rf-sections operating at opposite phase.

Both the single rf-field signal [S(0) + S(180)] and the inter-ference signal [S(0) - S(180)] of reduced width are shown in fig. 4. Total corrections of the data range from + 64 ppm (58 KHz) to + 149 ppm (136 KHz) depending on beam energy and the rf field spacing with an estimated uncertainty of ± 20 ppm (18 KHz). With the statistical un-certainty of ± 10 ppm the overall accuracy with respect to the Lamb shift interval is ± 19 ppm (20 KHz) corresponding to 1/5000 of the natural width of the 2 P state or about 1/1800 of the (reduced) signal width.

These values clearly indicate that for the $2 S_{1/2} - 2 P_{1/2}$ Lamb shift in H a further increase in experimental precision seems to be impossible.

In principle resonances on higher n states would have the advantage of narrower obtainable linewidth because of the relatively longer lifetime of n $P_{1/2}$ states for n > 2. The experiments reported for n = 3 and n = 4 states[26,27] in hydrogen and the n = 3, 4, 5 and 6 states in the helium ion[28-31] are less precise than the results obtained for n = 2 of H and D. This is mainly because the Lamb shift scales with quantum number n as $1/n^3$. In addition the lower populations achievable for n $S_{1/2}$ states (n > 2) and other technical difficulties somewhat limit the final precision quoted.

Table 1   Results for the 2 $S_{1/2}$ -2 $P_{1/2}$ Lamb shift in Hydrogen

Lamb shift	Reference

Experiment

1057.77 (06)	Triebwasser et al. (13)
1057.90 (06)	Robiscoe and Cosens (19)
1057.862 (020)	Andrews and Newton (24)
1057.893 (020)	Lundeen and Pipkin (25)

Theory

1057.910 (010)	Erickson (3,6)
1057.864 (014)	Mohr (7,32)

A comparison of the most precise experimental and theoretical values of the 2 $P_{1/2}$ - 2 $S_{1/2}$ Lamb shift of hydrogen is given in table 1 and is graphically displayed in fig. 5. The theoretical values obtained by Mohr (a,A)[7,32] and by Erickson (b,B)[3,6] differ by the rms proton radius r used for the calculation (a, b: r = 0.80 ± 0.02 f; A, B: r = 0.87 ± 0.02 f).[57] This 10 % variation in the proton radius causes a variation of the Lamb shift of the order of the accuracy of the best currently measured values. This strongly indicates that the Lamb shift measurements in H although extremely precise can no longer be considered as a very useful test for the QED unless more accurate values for the proton radius become available. This, however, can not be expected in the near future.

If the QED contributions are taken to be correct precision Lamb shift measurements could be used to extract information of the nuclear structure. In this way Andrews et al. derived an independent value of the proton radius of 0.845 ± 0.050 f.[33]

III   The Lamb shift of hydrogenic ions

In recent years, advances in techniques and instrumentation together with the use of nuclear heavy-ion accelerators have made possible an extension of Lamb shift experiments to hydrogenic ions of increasingly higher nuclear charge Z. In theory, new calculation techniques have been applied to high Z-systems. The obtained theoretical and experimental results have stimulated wide interest because the large enhancement that occurs in the Lamb shift due to an approximate Z^4 dependence allows sensitive studies of QED interactions in the high field regime. In addition the theoretical uncertainty due to the nuclear size uncertainty (which is about 50 % of the total calculated uncertainty for Z = 1) is considerably reduced for higher Z (it amounts to 20 % at Z = 6 and 13 % at Z = 17) and is no longer the limiting factor for a comparison between theory and experiment.[32]

It is impossible, however, to extrapolate the techniques used in the Lamb shift experiments for Z = 1 or 2 to higher Z hydrogenic systems. In order to remove Z - 1 electrons and populate relevant states, hot plasmas or high velocity beams must be used. In addition, in the region Z > 3 conventional rf-resonance absorption techniques would require very high magnetic fields and unreasonably high microwave power and frequency.

At present two approaches have been used to circumvent these difficulties: (a) stark quenching of the 2 $S_{1/2}$ metastable state with an applied field and (b) laser resonance absorption. Both methods require the use of beams of high-velocity highly ionized atoms.

The beam energy needed for hydrogenic ion production ranges from 6 MeV for Z = 5 to 25 MeV for Z = 10, but 170 MeV are necessary at Z = 15 and 28 GeV at Z = 80. For Lamb shift measurements it is desirable to maximize the hydrogenic charge-state fraction of the beam while minimizing the helium-like fraction, since the latter contributes to the radiative background in experiments which depend on detecting Lyman-α photons. This optimization may be accomplished in certain cases by directing a beam of fully stripped ions to the experimental area and then inducing one-electron pick-up to excited states via a beam-foil or beam-gas interaction. There is also evidence that electron-pickup processes enhance 2 $S_{1/2}$ state formation whereas higher angular momentum states are formed preferentially in stripping.

The method of stark quenching is based on the fact that stark coupling of the 2 $S_{1/2}$ and 2 $P_{1/2}$ via the electric dipole interaction in the presence of a DC electric field quenches the metastable state via an allowed E1 transition to the ground state. The transition rate is a function of the 2 S - 2 P energy difference.[34] This is the basis for an indirect measurement of the Lamb shift.

Including the mixing of the $2\,P_{3/2}$ and $2\,P_{1/2}$ states the decay rate γ of the $2\,S_{1/2}$ is given by

$$\gamma = \gamma_{2p}\,\frac{|V/h|^2}{\omega_L^2 + \frac{1}{4}\,\gamma_{2p}^2} + \frac{|M/h|^2}{\omega_F^2 + \frac{1}{4}\,\gamma_{2p}^2}$$

where

$$V = <2\,S_{1/2}\,|\hat{E}\,\hat{r}|\,2\,P_{1/2}> = \sqrt{3}\cdot e\ E\ a_0/Z$$

$$M = <2\,S_{1/2}\,|\hat{E}\,\hat{r}|\,2\,P_{3/2}> = \sqrt{6}\cdot e\ E\ a_0/Z$$

E is the applied electric field, a_0 is the Bohr radius, ω_L is the Lamb shift, ω_F is the $2\,P_{3/2} - 2\,S_{1/2}$ energy difference and γ_{2p} the decay rate of the 2 P state which can be calculated[48] in the dipole approximation to $\gamma_{2p} = 6.265 \times 10^8 Z^4$ sec^{-1}.

In most cases the $|M/h|^2$ term is relatively small. For $^{12}C^{5+}$, for example, it contributes by about 0.5 %. Higher order terms are still smaller by at least 10^{-2}.

For a measurement of the Lamb shift of $^{12}C^{5+}$ with the stark quenching method[35] a metastable beam was produced by first post-stripping the dominant C^{4+} component of 25 or 35 MeV beams to C^{6+}. The C^{6+} fraction was selected with an analyzing magnet. Then the bare carbon nuclei passed through a 0.1 Torr argon target for electron pickup. The target pressure had to be adjusted tc maximize the $2\,S_{1/2}$ production rate while maintaining a low enough helium-like metastable production. Some of these helium-like species contribute to background effects because their lifetimes and energy is comparable to the hydrogenic $2\,S_{1/2}$ state.

Leaving the target area the beam which now contains a C^{5+}, $2\,S_{1/2}$ component traversed a 2 m long drift region and entered the quenching area. As electric quenching field, the motional electric field experienced by the fast moving ions in a magnetic field transverse to the beam direction was used. A transverse magnetic field of 3 KG, for example, generated an effective electric field of about 60 KeV cm^{-1} which is sufficient to reduce the $^{12}C^{5+}$, $2\,S_{1/2}$ metastable decay length to about 2.8 cm at 25 MeV.

The 33.8 Å (367 eV) Lyman-α quenching radiation was monitored with a detector which is positioned perpendicular to the beam and can be moved parallel to the beam direction.

The detected counting rates as a function of the detector position are shown in figure 6. In this figure the change in slope proportional to B^2 is apparent whereas the square root dependence on energy is less obvious.

In order to extract a Lamb shift value from these Stark quenching data the particle velocity must be precisely known. Also the magnetic field strength must be measured very accurately. Without any corrections decay lengths from curves like those shown in fig. 5 are sufficient to determine the Lamb shift to about 2 %. In order to extract a value for S(n = 2) accurate to 1 % or better, all effects which might influence the result at the 0.1 % or greater level have to be considered. These effects include Zeeman level splitting due to the applied magnetic field, pre-quenching in the fringing field of the electromagnet, the proximity of the $2 P_{3/2}$ level, relativistic time dilation and beam deflection in the applied field.

For the C^{5+} ion the final result[36] was for the Lamb shift L (n = 2) = 780.1 ± 8.0 GHz. The resulting uncertainty contains the following contributions: 4 GHz statistics, 5 GHz due to background effects, 4 GHz geometrical effects because of changes in solid angle and segment width viewed by the detector because of the bending of the beam in the magnetic field, 2 GHz for a possible beam density non-uniformity and 1 GHz for field and energy determination.

A summary of experimental and theoretical results[6,32] as presently obtained for hydrogenic ions is given in table 2. A detailed review on these experiments has been given in ref. 37.

The helium data are in acceptable agreement with calculations and with each other, though a higher precision measurement would be desirable. Similar considerations are true for the Lithium values. The 1 % precision for the $^{12}C^{5+}$ Stark effect determination cannot distinguish the 0.2 % difference between the various calculations. The $^{16}O^{7+}$ Stark effect experiments are internally consistent and within quoted uncertainties, overlap the Lamb shift calculations. The measurements on $^{19}F^{8+}$ and $^{40}Ar^{17+}$ are the highest Z hydronic ions studied to date.

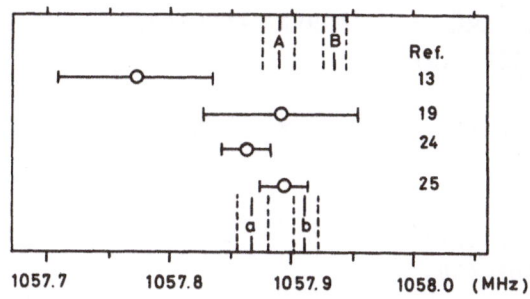

Figure 5   (From Ref. 8). Comparison of experimental and theoretical values of the 2 $S_{1/2}$ - 2 $P_{1/2}$ Lamb shift in hydrogen. For further details: see text.

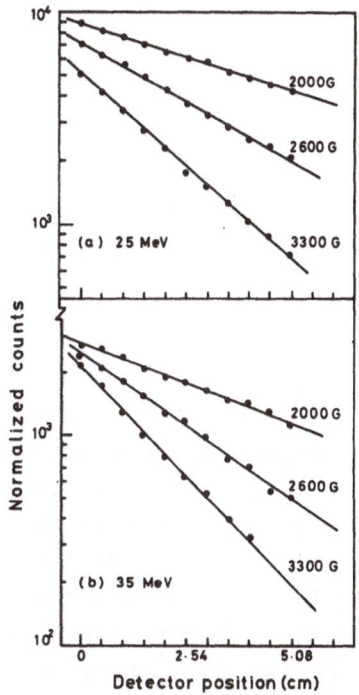

Figure 6   (From Ref. 35) Normalized $^{12}C^{5+}$ Lyman α counting rate plotted against detector position.

Table 2   Summary of experimental results for the Lamb shift in hydrogenic ions

System	Lamb shift (GHz)	Reference	Theory (Ref. 32 and 6)
$^4\text{He}^+$	14.0462 (12)	38	14.04205 (55)
	14.0402 (18)	39	14.04478 (610)
$^6\text{Li}^{2+}$	62.765 (21)	40	62.7375 (66)
	62.79 (07)	41	62.7620 (94)
	63.031 (327)	34	
$^{12}\text{C}^{5+}$	780.1 (8.0)	36	781.99 (21)
			783.68 (25)
$^{16}\text{O}^{7+}$	2215.6 (7.5)	42	2196.21 (92)
	2202.7 (11.0)	43	2205.2 (1.5)
$^{19}\text{F}^{8+}$	3339 (35)	44	3343.1 (1.6)
			3359.1 (3.0)
$^{40}\text{Ar}^{17+}$	38000 (600)	45	38250 (25)
			39039 (184)

Presently the $^{16}\text{O}^{7+}$ results are the most accurate Lamb shift measurements for a $Z > 3$ hydrogenic system with an experimental uncertainty of ± 0.5 %. More typical are uncertainties of about ± 1 % obtained in other experiments given in table 2. These results are at least two orders of magnitude less accurate than the latest Lamb shift results for $n = 2$ in H and D. Thus, for the future two trends are discernable: (1) using presently available techniques to study higher Z systems in order to amplify the Lamb shift and to render observable previously unmeasured higher order terms, and (2) improving experimental techniques in order to achieve sensitivities of the order of 0.1 to 0.01 %.

In experiments with ions of particularly high Z problems arise from the beam energy and intensity necessary to produce hydrogenic species in the required amounts. In addition the decrease in the $2\,S_{1/2}$ state lifetime[46,47] with increasing Z places limitations on the Lamb shift measurements which require the metastability of this state. At higher Z two photon electric dipole decay as well as single-photon magnetic dipole transitions to the ground state effectively reduce the 2 S lifetime which becomes less than 10^{-9} sec for $Z \geq 20$, for example.[46,47)]

Another main experimental difficulty in applying Stark quenching to increasingly higher Z is the bending of the beam due to the applied transverse field. This produces a position-dependent change in detector counting efficiency as the detector is moved along the beam axis. At higher Z the field strengths needed to produce a convenient detector count rate cause a position-dependent detector efficiency correction which becomes comparable to the total experimental uncertainty. These systematic effects may be avoided, however, with the anisotropy method which[49,50,51)] thus may become an important and precise tool. The principles of this method are described in detail in another contribution in this volume.

The improvements in experimental techniques to achieve sensitivi-ties of the order of 0.1 to 0.01 % will have useful applications in the study of hydrogenic systems currently available. Improved experimental sensitivities to below the 0.1 level will permit the observation of much higher order contributions and thus provide more stringent tests of present QED calculations.

A method which appears to offer presently good potential for achieving sensitivities of this order are laser resonance absorption in fast heavy ion beams. For Z > 5 the $2 P_{3/2} - 2 S_{1/2}$ and $2 S_{1/2} - 2 P_{1/2}$ transition energies occur in a spectral region that can be reached in many cases with current lasers. In principle the application of laser resonance absorption to the measurement of the Lamb shift in ions re-presents a return to the precision resonance techniques used for the Lamb shift measurements in H and D. However, a number of difficulties particular to fast, high Z ion experiments have arisen and have limited the experimental sensitivities of this technique, thus far, to about the same level as presently achievable with the Stark quenching method. Although these difficulties do not appear to be insurmountable the de-creasing $2 S_{1/2}$ metastability with increasing Z and the available laser power places limits on the general applicability of the laser resonance absorption technique.

Nevertheless, there exist several candidates for laser resonance absorption spectroscopy on hydrogenic ions.[37,52)] So far, the laser resonance method has been employed to measure the n = 2 Lamb shift in $^{19}F^{8+}$ using a doppler-tuned HBr laser.[44] With a tunable dye laser the energy difference in μ^-He^+ was determined.[53] In particular the very precise investigation in μ^-He^+ can be considered as an encouraging start for the application of laser absorption methods.

It should be mentioned that several Lamb shift experiments have been reported for the highly ionized helium-like ions $^{35}Cl^{15+}$ (ref.54), $^{16}O^{6+}$ (ref.55) and $^{40}Ar^{16+}$ (ref.56). In the helium-like systems, however, it will be extremely difficult to calculate atomic wavefunctions with sufficient accuracy to separate purely QED effects.

IV   Lamb shift of the 1 $S_{1/2}$ state

The Lamb shift experiments discussed so far did concentrate on the measurement of the 2 $S_{1/2}$ - 2 $P_{1/2}$ separation. Of similar interest is of course the shift of the (n = 1) $S_{1/2}$ ground state. Due to the $1/n^3$ dependence the Lamb shift of this level is 8 times larger than the shift of the n = 2 state.

In principle the Lamb shift of the 1 $S_{1/2}$ can be determined from the separation ΔE_{12} = 1 $S_{1/2}$ - 2 $P_{1/2}$ which corresponds to the energy of the L_α-radiation. For sufficiently high Z (in the range of Z~60-92) the self energy contribution to ΔE_{12} is on the order of 10^{-3} to 10^{-4} of the total L_α energy. Thus, using high resolution solid-state x-ray detectors or x-ray crystal spectrometers this QED contribution to the detected energy should be observable using current techniques.

For H and D, however, the Lamb shift of the 1 $S_{1/2}$ is only about $3 \cdot 10^{-6}$ of the ΔE_{12}-value. A precise measurement of this small deviation from the Dirac value of L_α is impossible for two reasons. First, the large Doppler shift of 30 GHz is almost 4-times larger than the Lamb shift contribution. Second, the Rydberg constant is known to 10^{-8}. Thus, the Dirac energy of the separation ΔE_{12} is known only to this accuracy.

These difficulties may be circumvented by comparing the separation ΔE_{12} with the energy difference ΔE_{24} = E(n = 4) - E(n = 2). ΔE_{12} is given by ΔE_{12} = R · 3/4 + C_{12}. C_{12} are corrections which include the Lamb shift contributions. ΔE_{24} equals R · 3/16 + C_{24}. The comparison of 1/4 · ΔE_{12} with ΔE_{24} provides the difference D = 1/4 · C_{12} - C_{24} of the corrections. Since the Dirac contributions to the value of D can be calculated a measurement of D can provide a value of the Lamb shift of the 1 S level.[59] Because of the comparison of ΔE_{12} and ΔE_{24} the precision of the value of the Rydberg is of no importance.

The energy separations ΔE_{12} and ΔE_{24} can be measured very precisely with modern dopplerfree nonlinear spectroscopy of two-photon excitation and saturated absorption spectroscopy. Such measurements, which have been described in detail in ref. 58,59 and 60 provide an easy method for a comparison of $1/4 \cdot \Delta E_{12}$ and ΔE_{24}. In these investigations[60] the 1 S Lamb shift of Deuterium was determined to 8177±30 MHz. The theoretical value is 8172.23 ± 0.12 MHz.[3]

The relatively large experimental error is mainly caused by corrections which are due to pressure and stark effects in the gas-discharge which is used for populating the 2 S state for the observation of the ΔE_{24} saturated absorption spectrum. In addition, experimental uncertainties arise in the ΔE_{12} two-photon spectrum due to Ac Stark effects and insufficient frequency control on the exciting pulsed, ultraviolet laser radiation.

These uncertainties should be reduceable by measuring the ΔE_{24} separation in a metastable atomic beam and exciting the 1 S - 2 S two-photon transition with continuous ultraviolet laser radiation of narrow bandwidth.

The ultimate resolution obtainable with these laser spectroscopic methods is limited only by the long lifetimes of the 2 S and 4 S states,[60] by the laser bandwidth, by the small second order doppler shift in the 1 S - 2 S spectrum (which amounts to about 30 KHz) and by a contribution of the ac Stark effect(about 50 to 80 KHz). Experimental linewidths as narrow as 10^5 Hz may be achievable with further advanced laser technology. The experimental accuracy which may be obtained in this way could well surpass the present theoretical precision. But as long as this value is limited by hadronic structure effects even such extremely precise experimental investigation will be of very limited value for a test of the QED.

References

1. W.E. Lamb, Jr., and R.C. Retherford, Phys. Rev. $\underline{72}$, 241 (1947).

2. B.E. Lautrup, A. Petermann, and E. de Rafael, Phys. Rep. $\underline{3}$, 193 (1972).

3. G.W. Erickson, J. Phys. Chem. Ref. Data $\underline{6}$, 831 (1977).

4. G.W. Erickson and D.R. Yennie, Ann. Phys. (N. Y.) $\underline{35}$, 271, 447 (1965).

5. T. Appelquist and S.J. Brodsky, Phys. Rev. $\underline{A2}$, 2293 (1970).

6. G.W. Erickson, Phys. Rev. Lett. $\underline{27}$, 780 (1971).

7. P.F. Mohr, Phys. Rev. Lett. $\underline{34}$, 1050 (1975).

8. H.J. Beyer, in Progress in Atomic Spectroscopy, Part A, W. Hanle and H. Kleinpoppen Eds., Plenum Press, New York, 1978.

9. W.E. Lamb, Jr. and R.C. Retherford, Phys. Rev. $\underline{79}$, 549 (1950).

10. W.E. Lamb, Jr. and R.C. Retherford, Phys. Rev. $\underline{81}$, 222 (1951).

11. W.E. Lamb, Jr., Phys. Rev. $\underline{85}$, 259 (1952).

12. W.E. Lamb, Jr. and R.C. Retherford, Phys. Rev. $\underline{86}$, 1014 (1952).

13. S. Triebwasser, E.S. Dayhoff, and W.E. Lamb, Jr., Phys. Rev. $\underline{89}$, 98 (1953).

14. E.S. Dayhoff, S. Triebwasser, and W.E. Lamb, Jr., Phys. Rev. $\underline{89}$, 106 (1953).

15. H.J. Beyer and H. Kleinpoppen in Progress in Atomic Spectroscopy, Part B, W. Hanle and H. Kleinpoppen Eds, Plenum Press, New York, 1978.

16. R.T. Robiscoe, Phys. Rev. $\underline{138}$, A 22 (1965).

17. R.T. Robiscoe and T.W. Shyn, Phys. Rev. Lett. $\underline{24}$, 559 (1970).

18. R.T. Robiscoe, pp. 373-376, Precision Measurements and Fundamental Constants, D.N. Langenberg and B.N. Taylor, NBS (USA), Special publ. No. 343 (1971).

19. R.T. Robiscoe and B.L. Cosens, Phys. Rev. Lett. $\underline{17}$, 69 (1966).

20. B.L. Cosens, Phys. Rev. $\underline{173}$, 49 (1968).

21. T.V. Vorburger and B.L. Cosens, pp. 361-365 of Ref.18 (1971).

22. D.A. Andrews and G. Newton, Phys. Rev. Lett. $\underline{37}$, 1254 (1976), G. Newton, D.A. Andrews, and P.J. Unsworth, Phil. Trans. R. Soc. London A $\underline{290}$, 373 (1979).

23. D.A. Andrews and G. Newton, J. Phys. **B8**, 1415, (1975).

24. D.A. Andrews and G. Newton, J. Phys. **B9**, 1453 (1976).

25. S.R. Lundeen and F.M. Pipkin, Phys. Rev. Lett. **34**, 1368 (1975).

26. C.W. Fabjan and F.M. Pipkin, Phys. Rev. **A6**, 556 (1972).

27. C.W. Fabjan, F.M. Pipkin, and M. Silverman, Phys. Rev. Lett. **26**, 347 (1971).

28. O. Mader, M. Leventhal, and W.E. Lamb, Jr., Phys. Rev. **A3**, 1832 (1971).

29. L. Hatfield and R. Hughes, Phys. Rev. **156**, 102 (1967).

30. R. Jacobs, K. Lea and W.E. Lamb, Bull. Am. Phys. Soc. **14**, 525 (1969).

31. A. Eibofner, Phys. Lett. **47A**, 399 (1974).

32. P.J. Mohr, Beam Foil Spectroscopy, I.A. Sellin and D.T. Pegg, eds., Plenum, New York, 1976; and 5th International Conference on Atomic Physics, Berkeley, 1976.

33. D.A. Andrews, R. Golub, and G. Newton, J. Phys. **G3**, L91 (1977).

34. C.Y. Fan, M. Garcia-Munoz, and I.A. Sellin, Phys. Rev. **161**, 6-15 (1967).

35. D.E. Murnick, M. Leventhal, and H.W. Kugel, Phys. Rev. Lett. **27**, 1625 (1971).

36. H.W. Kugel, M. Leventhal, and D.E. Murnick, Phys. Rev. **A6**, 1306 (1972).

37. H.W. Kugel and D.E. Murnick, Rep. Prog. Phys. **40**, 297 (1977).

38. M. Narasimham and R. Strombotne, Phys. Rev. **A4**, 14 (1971).

39. E. Lipworth and R. Novick, Phys. Rev. **108**, 1434 (1957).

40. M. Leventhal, Phys. Rev. **A11**, 427 (1975).

41. D.D. Dietrich, P. Lebow, R. de Zafra, and H. Metcalf, Bull. Am. Phys. Soc. **21**, 625.

42. G.P. Lawrence, C.Y. Fan and S. Bashkin, Phys. Rev. **28**, 1612 (1972)

43. M. Leventhal, D.E. Murnick, and H.W. Kugel, Phys. Rev. Lett. **28**, 1609 (1972).

44. H.W. Kugel, M. Leventhal, D.E. Murnick, C.K.N. Patel, and O.R. Wood II, Phys. Rev. Lett. **35**, 647 (1975).

45. H. Gould and R. Marrus, Phys. Rev. Lett. **41**, 1457 (1978).

46. S. Klarsfeld, Phys. Lett. 30A, 382 (1969).

47. W.R. Johnson, Phys. Rev. Lett. 29, 1123 (1972).

48. H.A. Bethe and E.E. Salpeter, Handb. Phys. XXXV 371.

49. G.W.F. Drake and R.B. Grimley, Phys. Rev. A8, 157 (1973).

50. G.W.F. Drake and Lin Chien-Ping, Phys. Rev. A14, 1296 (1976).

51. G.W.F. Drake, P.S. Farago, and A. van Wijngaarden, Phys. Rev. A11, 1621 (1975).

52. D.E. Murnick, in Beam Foil Spectroscopy, I.A. Sellin and D.T. Pegg eds., (New York: Plenum), 1976.

53. A. Bertin et al., Phys. Lett. 55B, 41 (1975).

54. H.G. Berry, R. De Serio, and A.E. Livingston, Phys. Rev. Lett. 41, 1652 (1978).

55. N.A. Jelley, I.A. Armour, S. Bashkin, E.G. Myers, R. O'Brien and J.D. Silver, J. Phys. B. 12, 2605 (1979).

56. W.A. Davis and R. Marrus, Phys. Rev. A15, 1963 (1977).

57. F. Borkowski, G.G. Simon, V.H. Walther, and R.D. Wendling, Z. Phys. A275, 29 (1975).

58. T.W. Hänsch, S.A. Lee, R. Wallenstein, and C. Wieman, Phys. Rev. Lett. 34, 307 (1975)

59. S.A. Lee, R. Wallenstein, and T.W. Hänsch, Phys. Rev. Lett. 35, 1262 (1975).

60. C. Wieman and T.W. Hänsch, Phys. Rev. A22, 193 (1980).

HELIUM AND HELIUM-LIKE SYSTEMS

R. Neumann

Physikalisches Institut der Universität Heidelberg
Germany

The two-electron system helium and the members of its isoelec-
tronic sequence belong to the fundamental problems in atomic physics.
Their spectra are most important with respect to very accurate
calculations of quantum-electrodynamic, relativistic and nuclear
structure corrections. These calculations became feasible with the
advent of the large electronic computers. The subject of two-electron
atoms has many ramifications, even under the limiting aspect of
quantum electrodynamics. This report will concentrate on experimental
fine and hyperfine structure investigations of the lowest S and P
states in He and Li^+. Clearly this choice does by no means intend to
impair the importance of the manifold activities concerning He-like
systems which are not treated or cited here.

Many calculations and experiments dealt specially with He and
aspired to highest accuracy. In the first part of this lecture two
high-precision measurements in He will be outlined: a recent investigation of the $1s2p\ ^3P$ fine structure (fs) splitting of ^4He (1) and a
measurement of the $1s2s\ ^3S_1$ hyperfine structure (hfs) in ^3He (2).
Secondly laser saturation spectroscopy in the $2\ ^3P$ state of $^{6,7}Li^+$
(3), and a combined laser microwave measurement of the $2\ ^3S_1$ hfs of
$^7Li^+$ will be described (4). Finally the question of future precision
measurements in the next He-like ions from Be^{2+} to F^{7+} is to be
briefly discussed.

The quantum-electrodynamic theory of the two-electron atoms
(5,6,7) can be tested most sensitively with the fs splitting of the
$1s2p\ ^3P$ state of ^4He. Measurements of each of the two splittings have
been performed some years ago (8,9,10) and a refined calculation was

presented (11). A recent experiment which is incidentally the most accurate fs measurement yet made, measured the sum of the two splittings with a one-photon microwave transition (1)*.

   A beam of He atoms excited to the metastable 1s2s $^3S_1$ state by electron impact, passes a microwave cavity. The atoms are excited with a lamp to the 2 $^3P_{0,1,2}$ multiplet where a one-photon magnetic dipole transition from J=2,M_J=0 to J=0,M_J=0 takes place in the microwave field of fixed frequency. This transition which is forbidden in zero magnetic field, is induced in a field of 2 kG where J is no longer a good quantum number. The respective part of the ^{4}He level diagram is shown in Fig.1 and the main features of the interaction region are sketched in Fig.2.

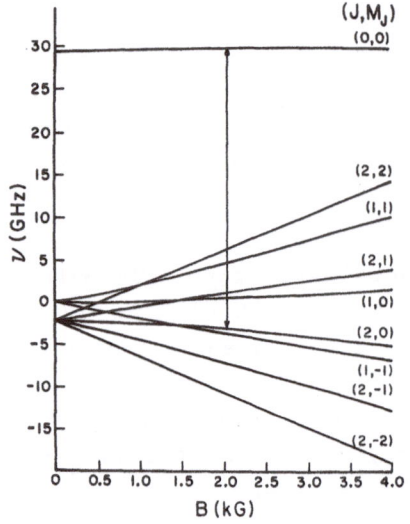

<u>Fig.1:</u> Zeeman level scheme of the 1s2p ^{3}P state of ^{4}He (taken from Ref.1)

---

* After this manuscript was completed the author received two further preprints of extensive papers concerning ^{4}He, submitted to Phys.Rev.A Aug 1, 1980. They are cited as Ref.39 and 40.

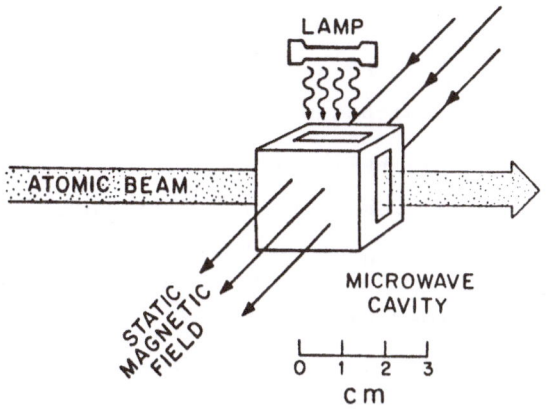

Fig.2: Schematic diagram of the main features of the interaction region (taken from Ref.1)

The 2 ^{3}P term has a lifetime of 10^{-7} sec and decays back to the 2 ^{3}S$_1$ state. Caused by the RF resonance a partial 2 ^{3}S$_1$ M$_J$ sublevel transfer happens and is monitored by magnetic deflection. A signal curve plotted as a function of the magnetic field (calibrated with NMR) is given in Fig.3.

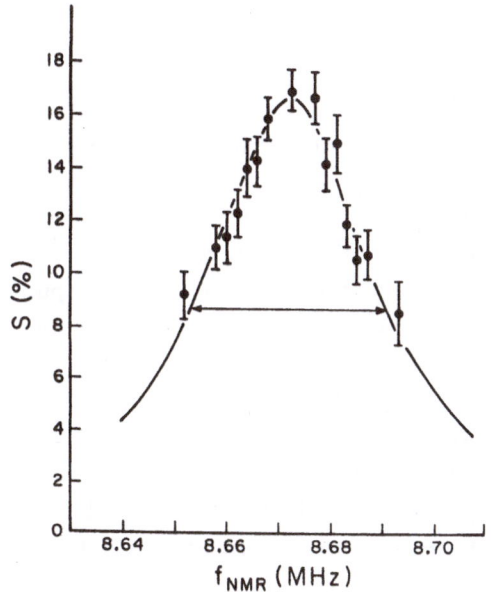

Fig.3: Resonance signal as a function of magnetic field measured in units of proton NMR. Error bars are 1 standard deviation. Solid curve is a least-squares fit. (taken from Ref.1)

The experimental result is

$$\nu(P_0-P_2) = 31.908040 \ (20) \ \text{GHz} \ (0.6 \ \text{ppm}).$$

Subtracting an earlier measurement (9)

$$\nu(P_1-P_2) = 2.291196 \ (5) \ \text{GHz} \quad (2.2 \ \text{ppm})$$

gives an improved value

$$\nu(P_0-P_1) = 29.616844 \ (21) \ \text{GHz} \ (0.7 \ \text{ppm}),$$

in comparison to an older less precise direct measurement

$$\nu(P_0-P_1) = 29.616864 \ (36) \ \text{GHz} \ (1.2 \ \text{ppm}). \quad (10)$$

This procedure is done in order to compare with the best theoretical value (11) which is available for the (P_0-P_1) transition. This value together with the theoretical contributions calculated so far is given in Table 1. The calculation was performed with

$$\alpha^{-1} = 137.035963 \ (15) \quad (0.11 \ \text{ppm})$$

and with

$$Ry = 109737.31476 \ (32) \ \text{cm}^{-1} \quad (0.0029 \ \text{ppm})$$

taken from (12) and (13) respectively. Additionally the theory was used to extract a value of the fine structure constant from the experimental result. The authors give

$$\alpha^{-1} = 137.03613 \ (11) \quad (0.8 \ \text{ppm}).$$

A third recent value of

$$\alpha^{-1} = 137.036006 \ (11) \quad (0.08 \ \text{ppm})$$

is given by (14). Many other investigations concerned the term structure of higher states in He (15-20).

The second experiment described here (2) concerns the $2 \ ^3S_1$ hfs splitting of ^{3}He (nuclear spin I = 1/2) measured by the optical pumping magnetic resonance method. Fig.4 shows the respective terms of ^{3}He while the experimental layout is drawn schematically in Fig.5. The hfs of a hydrogenic atomic level with quantum numbers nl goes as $n^{-3}(2l+1)^{-1}$. Therefore the 2s electron is neglected in the lowest order approximation $\Delta\nu_0$ of the $2 \ ^3S_1$ hfs splitting. Table 2 summarizes the numerous necessary corrections to $\Delta\nu_0$ and is taken from (2). It should be emphasized that the QED corrections amount to nearly 1000 ppm. On the other hand there are large nuclear structure corrections an estimate of which is much more difficult.

The experimental result is

$$\Delta\nu(^3\text{He},2 \ ^3S_1) = 6 \ 739 \ 701 \ 177 \pm 16 \ \text{Hz} \quad (2.4\cdot10^{-3} \ \text{ppm}).$$

It is believed that the ratio

$$\rho = \Delta\nu(^3\text{He},2 \ ^3S_1)/\Delta\nu(^3\text{He}^+,2 \ ^2S_{1/2})$$

Table 1: Theoretical contributions to the fine structure of 2 ^{3}P in MHz (taken from Ref.1)

α^2Ry	m/Mα^2Ry	α^3Ry	α^4Ry	total theory
29564.587	-10.707	54.708	8.326	29616.914
±0.009	±0.00044		±0.042	±0.043 (1.5 ppm)

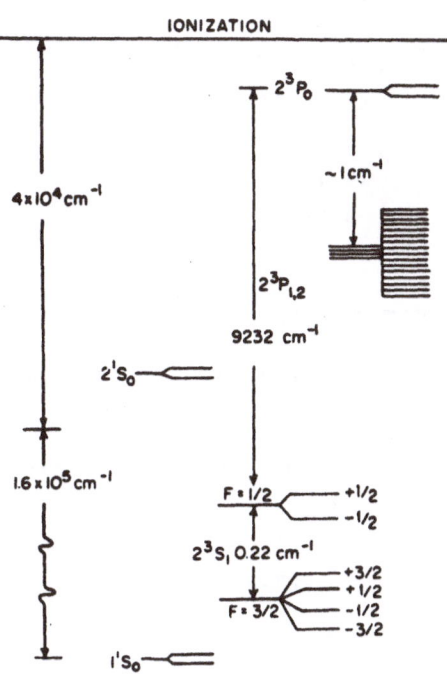

Fig.4: Respective part of the ^{3}He energy diagram (taken from Ref.2)

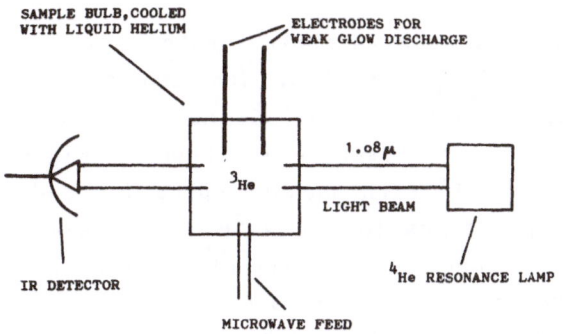

Fig.5: ^{3}He experiment

Table 2: A summary of the corrections to the hyperfine structure of
^{3}He in the 2 ^{3}S$_1$ state

Origin	Magnitude (ppm)
Presence of second electron	37000
Reduced mass	-550
Second-order contribution	-8.89
Use of relativistic wave functions	315
QED corrections	948.2
Relativistic reduced mass	-29
Spin hfs in adiabatic approximation	-146 → -183
Nuclear D state	3.3
Correction to adiabatic approximation	11.7
Nuclear orbital motion	0.8
Proton and neutron structure	-13
Diamagnetic shielding	4.3

is independent of nuclear structure effects and tests the QED corrections to hfs. With

$$\Delta\nu_{exp}(^3He^+,2\ ^2S_{1/2}) = 1083.3549807\ (88)\ MHz\ \ (8.1\cdot10^{-3}\ ppm)$$

taken from (21) follows

$$\rho_{exp} = 6.2211381\ (5.2\cdot10^{-8})\ \ (8.45\cdot10^{-3}\ ppm).$$

Comparison with

$$\rho_{theor} = 6.2211157\ (187)$$

shows good agreement. For an earlier precise measurement of $\Delta\nu_{exp}$ see
(22).

The following section is dedicated to measurements in Li$^+$, concerning again the 2 ^{3}S$_1$ and 2 ^{3}P$_{0,1,2}$ states. The two states are
connected via a resonant transition with λ = 5485 Å which is ideal for
dye laser spectroscopy. The energy level scheme is shown in Fig.6 for
the two stable isotopes ^{6}Li$^+$ and ^{7}Li$^+$ with nuclear spins I=1 and 3/2.
Both isotopes show large hfs splittings in the 2 ^{3}P$_{1,2}$ fs terms, and
hfs sublevels with equal quantum number F in different J levels perturb
each other significantly. Therefore J is not a good quantum number. In
order to extract the unperturbed fs splittings and hfs constants, all
the 2 ^{3}P splittings of the respective isotope must be measured and the
energy matrix has to be diagonalized. This illustrates that the situa-

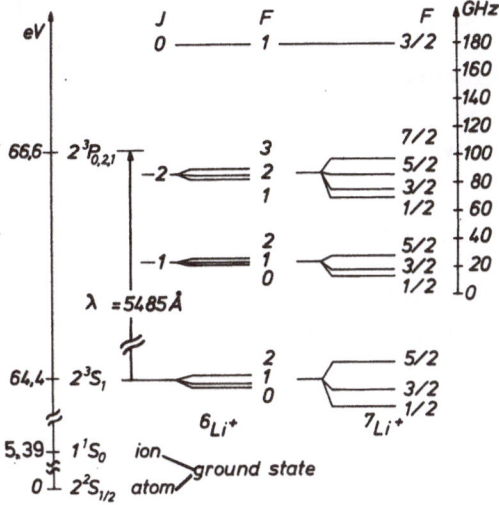

Fig.6: Energy diagram of the $2\,^3S_1$ and $2\,^3P$ terms of $^6Li^+$ and $^7Li^+$, the isotope shift is omitted

(taken from Ref.3)

tion is more awkward than it is for $2\,^3P$ fs investigations in ^4He besides the fact that one deals with ions rather than atoms and with much larger fs splittings, not easily accessible to microwave spectroscopy.

The fine and hyperfine components of the $(2\,^3S_1 - 2\,^3P)$ resonance line of $^{6,7}Li^+$ are spread over a frequency scale of about 200 GHz. With dye laser saturation spectroscopy all these splittings had been measured (3) and were calibrated with a stabilized confocal Fabry-Perot interferometer. Fig.7 shows a single laser scan over more than 50 GHz crossing the largest gap in the mixed spectrum of the two isotopes.

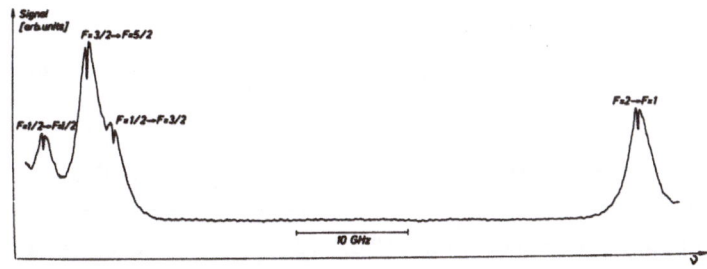

Fig.7: Signals of the $2\,^3S_1(F=2) - 2\,^3P_0(F=1)$ transition in $^6Li^+$ and $2\,^3S_1 - 2\,^3P_2$ transitions in $^7Li^+$ measured with a single laser scan over more than 50 GHz.
(taken from Ref.3)

Table 3 comprises the set of experimental fs splittings and hfs constants of the 2 ^{3}P state of ^{6,7}Li$^+$ together with theoretical values (7,23). Besides the many experimental and theoretical publications dealing with Li$^+$ and cited in (3), a recent measurement in the same transition with Doppler-tuned dye laser spectroscopy has to be mentioned (24).

In order to get independent of the imponderables of a calibration interferometer, measurements with laser optical pumping and microwave transitions were started. The first result is a presicion measurement of the 2 ^{3}S$_1$ hfs of ^{7}Li$^+$ (4).

Table 3: Hfs constants and fs splittings of ^{6}Li$^+$ and ^{7}Li$^+$, all values in MHz. (taken from Ref.3)

^{6}Li$^+$

	Ref.3	Ref.23,Ref.7(*)
A_c	1 390(6)	1 392.8
A_o	20.8(4.0)	18.4
A_d	-4.2(1.0)	-3.7
ν_{01}	-155 698(20)	-155 725.1*
ν_{02}	-93 023(9)	-93 072.1*

^{7}Li$^+$

	Ref.3	Ref.23,Ref.7(*)
A_c	3 669(6)	3 678.4
A_o	57(5)	48.6
A_d	-11.5(1.0)	-9.86
ν_{01}	-155 694(24)	-155 725.1*
ν_{02}	-93 019(7)	-93 072.1*

The method was used before for molecules (25), atoms (26) and ions (27). The experimental set-up is shown in Fig.8a and b. ^{7}Li$^+$ ions are produced and excited to the 2 ^{3}S$_1$ term (lifetime $\tau \approx$ 50 sec (28)) by electron bombardment right at the aperture of an oven, and formed to a well collimated beam with a kinetic energy of typically 200 eV ($v \approx 7.4 \cdot 10^6$ cm·sec^{-1}). A single-frequency dye laser light beam crosses the ion beam and depletes the population of one of the three 2 ^{3}S$_1$ hfs sublevels by optical pumping via one of the 2 ^{3}P hyperfine

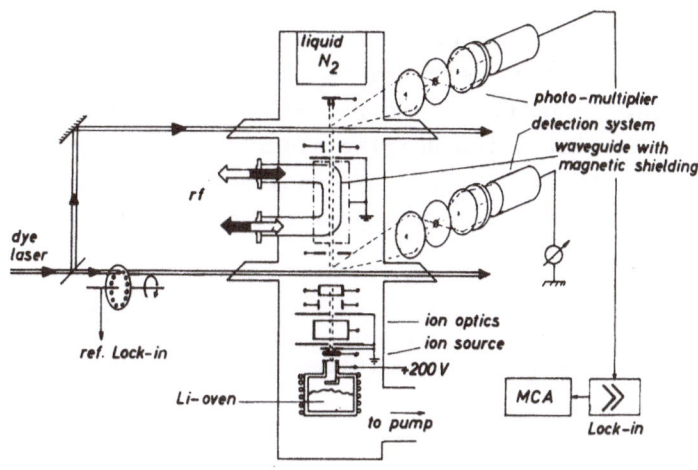

MICROWAVE-SYSTEM

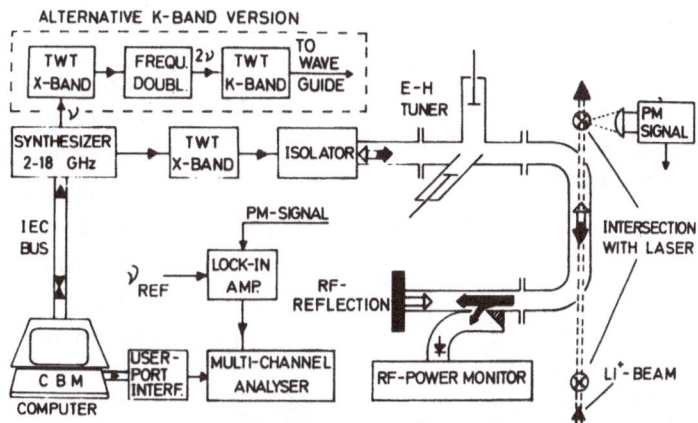

<u>Fig.8a and b:</u> Laser-microwave spectrometer and Li⁺ ion beam apparatus
(a), and microwave system in more detail (b).

states. The $2\,^3P$ lifetime is 43 nsec (29). The ion beam then passes a
waveguide of about 9 cm length where RF transitions within the $^3S_1$ term
are induced. The microwave radiation equalizes the population between
two neighbouring levels, and this is monitored via the change of
fluoresecence light intensity with a second laser beam crossing the ion
beam. The microwave region is magnetically shielded, and the earth
magnetic field is reduced to about 10 mG. Fig.8b gives the microwave
system in more detail with the alternative X-band and K-band versions
for the $F=1/2$ - $F=3/2$ ($\sim$12 GHz) and $F=3/2$ - $F=5/2$ ($\sim$20 GHz) transitions
respectively. A microcomputer varies the synthesizer frequency in

20 kHz steps and switches the channel number of the multichannel analyser. Since the microwave is reflected at the end of the waveguide, so that one wave travels with the ion motion and the other wave in opposite direction, two Doppler signals arise shifted by $\pm v_0 \cdot v/v_p$ where v_0 is the unshifted centre frequency and v_p is the RF phase velocity in the waveguide. Two typical signal curves are shown in Fig.9a and b.

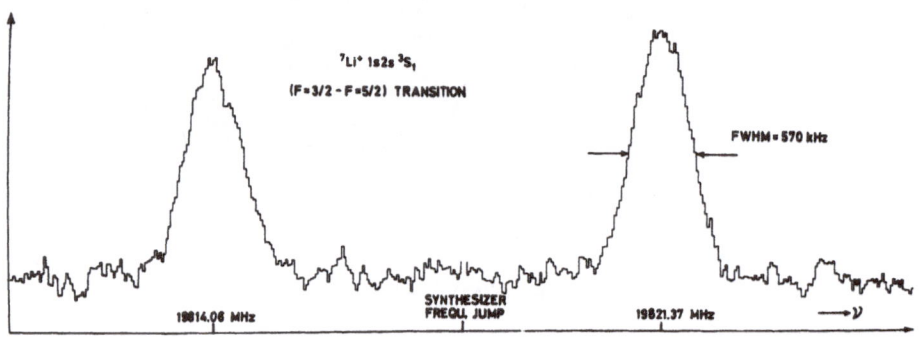

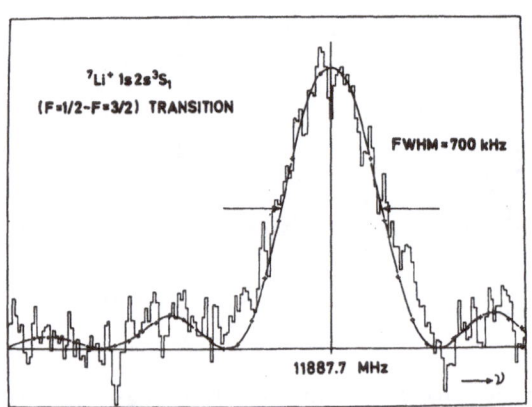

Fig.9a and b: $2\,^3S_1$, F=3/2-F=5/2 (a) and F=1/2-F=3/2 (b) microwave transitions. Both Doppler-shifted signals are given in (a). A theoretical fit curve is shown additionally in b.

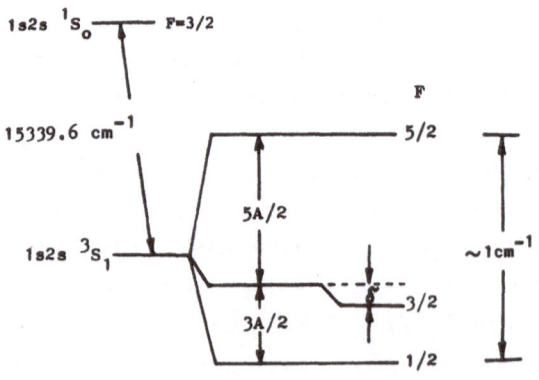

Fig.10: Scheme of the $2\,^3S_1$ hyperfine multiplet of $^7Li^+$ with the F=3/2 depression.

<u>Table 4:</u> Experimental values of the 2 $^3S_1$ hfs splitting*

F=1/2–F=3/2 splittings	11890.048 MHz ± 15 kHz
F=3/2–F=5/2 splittings	19817.703 MHz ± 15 kHz
Magnetic hfs constant A	7926.938 MHz ± 5 kHz
F=3/2 depression	359 kHz ± 11 kHz

* The given values should be regarded as preliminary. A detailed analysis of the results and a comprehensive description of the whole experiment will be given in (41).

The 2 $^3S_1$ hfs splitting can be expressed with the magnetic hfs constant A (see Fig.10). Mixing of the 2 $^3S_1$ term with the 2 $^1S_0$, with the latter represented by the nuclear spin quantum number I=3/2 causes a selective depression of the F=3/2 sublevel (3). Table 4 summarizes the experimental results.

Calculation e.g. of the (F=1/2–F=3/2) splitting with corrections only for the second electron (31), the anomalous magnetic moment in lowest order and for reduced mass (5) gives 11887.7(3) MHz. The error is dominated by the uncertainty of the second-electron correction. The difference to the experimental value is due to higher order radiative and relativistic corrections and probably to a significant nuclear structure influence. Similar measurements of the 2 $^3S_1$ hfs splittings in $^6Li^+$ are in progress. The aim is to check whether there is a hyperfine anomaly (32).*

Finally a brief outlook will be given on the next members of the iso-electronic sequence following Li^+, namely Be^{2+}, B^{3+}, C^{4+}, N^{5+}, O^{6+} and F^{7+}. Tables 5 and 6 summarize some characteristics of these elements such as the stable isotopes and there abundance in the natural isotopic mixture, the nuclear spin and some basic features of the ionic key terms 2 $^3S_1$ and 2 ^3P.

While the energy distance of the 2 $^3S_1$ state from the ion ground state is exceedingly large, there is a difference of only a few eV between 2 $^3S_1$ and 2 ^3P and the transition wavelength decreases slowly from the near ultraviolet in Be^{2+} to little less than 1500 Å in F^{7+}. For Be^{2+} and B^{3+} the transition wavelengths can be supplied with a power of several mW by cw dye laser frequency doubling, and in

* For the hyperfine anomaly in the Li atom see (42).

Table 5: Some properties of Be^{2+}, B^{3+}, and C^{4+}

	Be^{2+}	B^{3+}	C^{4+}	
stable isotopes	^{9}Be 100%	^{10}B 19.78% ^{11}B 80.22%	^{12}C 98.89% ^{13}C 1.11%	*
nuclear spin I	3/2	3 (^{10}B) 3/2 (^{11}B)	0 (^{12}C) 1/2 (^{13}C)	*
2 $^{3}S_{1}$ excitation voltage from atom ground state	146 V	270 V	447 V	**
wavelength 2 $^{3}S_{1}$ – 2 ^{3}P	3722 Å	2823 Å	2274 Å	**
fs splittings J=0 – J=1 J=1 – J=2	347 GHz 446	486 GHz 1578	375 GHz 4069	#
^{3}P lifetime	29 nsec	22 nsec	18 nsec	##
^{3}P natural term width	5.5 MHz	7.3 MHz	9 MHz	##

* Handbook of Chemistry and Physics, 55th edition, 1974/75
** C.E. Moore: Atomic Energy Levels, vol.I, Nat.Stand.Ref. Data Ser., Nat.Bur.Stand.(U.S.), 1971
# B. Schiff, Y. Akkad, C.L. Pekeris: Phys.Rev.A8, 2272 (1973)
W.L. Wiese, M.W. Smith, B.M. Glennon: Atomic Transition Probabilities, vol.I, Nat.Stand.Ref. Data Series, Nat.Bur.Stand.(U.S.), 1966

principle also the C^{4+} transition wavelength of 2274 Å should soon be feasible by frequency summing or doubling. Thus laser spectroscopy of the 2 $^{3}S_{1}$ and 2 ^{3}P states of Be^{2+}, B^{3+} and C^{4+} can be realized in the near future and would allow precision measurements of the fs, hfs, isotope shift (except for Be^{2+}) and Lamb shift.

E.g. the J=1 – J=2 fs splitting grows from about 60 GHz in Li^{+} to $\sim$4070 GHz in C^{4+} whereas the 2 ^{3}P lifetime decreases slowly from 43 to 17.7 nsec, corresponding to linewidths of 3.6 to 9 MHz. In the present there is no way to attack the large fs splittings with microwave technique. Instead of that a low velocity ion beam prepared in the 2 $^{3}S_{1}$ metastable state could be crossed at right angle with a frequency-doubled cw dye laser beam in single mode operation. Absolute wave-

Table 6: Some properties of N^{5+}, O^{6+}, and F^{7+}

	N^{5+}	O^{6+}	F^{7+}	
stable isotopes	^{14}N 99.63% ^{15}N 0.37%	^{16}O 99.76% ^{17}O 0.04% ^{18}O 0.2 %	^{19}F 100%	*
nuclear spin I	1 (^{14}N) 1/2 (^{15}N)	0 (^{16}O) 5/2 (^{17}O) 0 (^{18}O)	1/2	*
2 $^{3}S_1$ excitation voltage from atom ground state	687 V	994 V	1382 V	**
wavelength 2 $^{3}S_1$ - 2 ^{3}P	1902 Å	1630 Å	1423 Å	**
fs splittings J=0 - J=1 J=1 - J=2	260 GHz 8716	1763 GHz 16512	4531 GHz 28658	#
^{3}P lifetime	14.8 nsec	12.6 nsec	10.9 nsec	##
^{3}P natural term width	10.8 MHz	12.6 MHz	14.6 MHz	##

* etc. see Table 5

length measurements of the transition from the 2 $^{3}S_1$ state to the different 2 ^{3}P fs terms can be performed e.g. with a travelling Michelson interferometer (33 and references cited therein) to 10^{-8} or better, if the laser is stabilized to the saturation dip of the respective transition. Thus the fs splitting of C^{4+} could be achieved with an accuracy of about $5 \cdot 10^{-6}$.

Especially C^{4+} is an interesting case since $^{12}C^{4+}$ is the first He-like system after ^{4}He without nuclear spin so that a 2 ^{3}P fs measurement would be free of hfs perturbation. Additionally $^{13}C^{4+}$ is the next iso-electronic ion with nuclear spin 1/2 after ^{3}He with an estimated 2 $^{3}S_1$ hfs splitting of about 60 GHz.

The (2 $^{3}S_1$ - 2 ^{3}P) transition wavelengths of N^{5+}, O^{6+}, and F^{7+} are less than 2000 Å, preventing frequency doubled cw dye laser spectroscopy in the near future. In the case of F^{7+} the J=1-J=2 fs splitting fits with a CO_2 laser line at 10.6µ (34).

Among the subjects not discussed in this report is the storage of ions in a trap which is an alternative way to study the spectra of He-like ions (21,28). This naturally opens the possibility of ultimate precision. Secondly the high Z two-electron systems should be mentioned since they are especially suitable for QED tests via Lamb-shift and fine structure investigations (35 - 38).

The author is greatly indebted to Professor G. zu Putlitz and Dr. J. Kowalski for helpful discussions and advice.

The Li^+ experiment is sponsored by the Deutsche Forschungsgemeinschaft.

References:

1. W.E. Frieze, E.A. Hinds, V.W. Hughes, F.M. Pichanick: Phys.Lett. 78A, 322 (1980)

2. S.D. Rosner, F.M. Pipkin: Phys.Rev.A 1, 571 (1970)

3. R. Bayer, J. Kowalski, R. Neumann, S. Noehte, H. Suhr, K. Winkler, G. zu Putlitz: Z.Physik A 292, 329 (1979)

4. U. Kötz, J. Kowalski, R. Neumann, S. Noehte, H. Suhr, K. Winkler, G. zu Putlitz: 7th International Conf. on Atomic Physics, Boston, 4-8 Aug 1980, Abstracts p.162

5. H.A. Bethe and E.E. Salpeter: Quantum Mechanics of One- and Two-Electron Atoms. Berlin, Göttingen, Heidelberg: Springer 1957

6. M.L. Lewis, in Atomic Physics, edited by G. zu Putlitz, E.W. Weber and A. Winnacker (Plenum, New York, 1975), vol.4, p.105

7. Y. Akkad, C.L. Pekeris, B. Schiff: Phys.Rev. A 4, 516 (1971)

8. F.M.J. Pichanick, R.D. Swift, C.E. Johnson, V.W. Hughes: Phys.Rev. 169, 55 (1968)

9. S.A. Lewis, F.M.J. Pichanick, V.W. Hughes: Phys.Rev.A 2, 86 (1970)

10. A. Kponou, V.W. Hughes, C.E. Johnson, S.A. Lewis, F.M.J. Pichanick: Phys.Rev.Lett. 26, 1613 (1971)

11. M.L. Lewis, P.H. Serafino: Phys.Rev.A 18, 867 (1978)

12. E.R. Williams, P.T. Olsen: Phys.Rev.Lett. 42, 1575 (1979)

13. J.E.M. Goldsmith, E.W. Weber, T.W. Hänsch: Phys.Rev.Lett. 41, 1525 (1978)

14. H. Dehmelt, T. Kinoshita: private communication

15. K.B. MacAdam, W.H. Wing: Phys.Rev.A 15, 678 (1977)

16. J.W. Farley, K.B. MacAdam, W.H. Wing: Phys.Rev.A 20, 1754 (1979)

17. P.B. Kramer, F.M. Pipkin: Phys.Rev.A 18, 212 (1978)

18. E. Giacobino, E. De Clercq, F. Biraben, G. Grynberg, B. Cagnac: in: Laser Spectroscopy IV, Proceedings of the Fourth International Conference, Rottach Egern, 1979. H. Walther, K.W. Rothe (eds.). Berlin, Heidelberg, New York: Springer 1979

19. R. Panock, M. Rosenbluh, B. Lax, T.A. Miller: Symposium on Atomic Spectroscopy (SAS-79), Tucson, Arizona, Sep 10-14, 1979, Abstracts, p.52

20. G. von Oppen, S. Aynacioglu, W.-D. Perschmann, D. Szostak, A. Wolf: SAS-79, Abstracts, p.8

21. M.H. Prior, E.C. Wang: Phys.Rev.A 16, 6 (1977)

22. R. Novick, E.D. Commins: Phys.Rev.111, 822 (1958)

23. A.N. Jette, T. Lee, T.P. Das: Phys.Rev.A 9, 2337 (1974)

24. R.A. Holt, S.D. Rosner, T.D. Gaily, A.G. Adam: Phys.Rev.A 22, 1563 (1980)

25. S.D. Rosner, R.A. Holt, T.D. Gaily: Phys.Rev.Lett. 35, 785 (1975)

26. W. Ertmer, B. Hofer: Z.Physik A 267, 9 (1976)

27. S.D. Rosner, T.D. Gaily, R.A. Holt: Phys.Rev.Lett. 40, 851 (1978)

28. R.D. Knight, M.H. Prior: Phys.Rev.A 21, 179 (1980)

29. H. Schmoranzer, D. Schulze-Hagenest, S.A. Kandela: SAS-79, Abstracts, p.195

30. M.M. Sternheim: Phys.Rev.Lett. 15, 545 (1965)

31. P.J. Luke, R.E. Meyerott, W.W. Clendenin: Phys.Rev. 85, 401 (1952)

32. H.M. Foley: in: Atomic Physics, p.509, Proceedings of the First International Conference on Atomic Physics, Plenum Press 1969

33. J. Cachenaut, C. Man, P. Cerez, A. Brillet, F. Stoeckel, A. Jourdan, F. Hartmann: Revue de Physique Appliquée 14, 685(1979)

34. H.J. Andrä, J. Macek, J. Silver, N. Jelley, L.C. McIntyre: in: Beam-Foil Spectroscopy, vol.2. New York, Plenum Press (1976)

35. A.M. Ermolaev: in: Progress in Atomic Spectroscopy, Part A, New York and London, Plenum Press 1978

36. H.G. Berry, R. DeSerio, A.E. Livingston: Phys.Rev.Lett. 41, 1642 (1978)

37. R. DeSerio, H.G. Berry, A.E. Livingston: SAS-79, Abstracts, p.101

38. J.D. Silver: SAS-79, Abstracts, p.2

39. A. Kponou, V.W. Hughes, C.E. Johnson, S.A. Lewis, F.M.J. Pichanick: Submitted to Phys.Rev.A Aug 1, 1980

40. W. Frieze, E.A. Hinds, V.W. Hughes, F.M.J. Pichanick: Submitted to Phys.Rev.A Aug 1, 1980

41. U. Kötz, J. Kowalski, R. Neumann, S. Noehte, H. Suhr, K. Winkler, G. zu Putlitz: to appear in Z.Physik A 300 (1981)

42. E. Arimondo, M. Inguscio, P. Violino: Rev.Mod.Phys.49, 31, (1977)

Comparison Between Experiment and Theory in Heavy Electronic Systems

B. Fricke

Fachbereich Physik, Gesamthochschule Kassel, D-3500 Kassel, W. Germany

Introduction

Most talks in this symposium dealt with electronic and muonic systems with small or medium Z and one or two electrons or one muon, respectively. The quantum-electrodynamical effects in these systems are relatively small, but due to the very accurate measurements one is able to study them up to very high orders. On the other hand, the talk of Dr. Rafelski dealt with extreme electronic systems with Z around 170, where the QED effects are expected to be relatively big, but also relatively inaccurate from a computational point of view. Although these are systems with very many electrons, they have been treated there as one-electron systems, first, because most electrons are outer electrons, and thus do not play any important role and second, because the influence of the few other inner electrons does not change the predictions qualitatively, which are mainly connected with the question of the diving of the 1s level into the negative continuum. I would like to discuss here the area between these two extremes. These are systems

    a) with large and very large Z, where
    b) the many-body effects become important, and
    c) the observable effects are neither small nor big.

This area is the region of the binding energies of the innermost electrons of very heavy atoms (Z > 80).

The experimental data in this region result either from photo-electron spectroscopy[1] with an accuracy of the order of eV at binding energies of about 100 keV or from the observation of normal X-rays with an accuracy which is already below 1 eV [2].

The main assumption in every theoretical discussion of a many-electron system is an extremely good knowledge of the self-consistent field solution of the many-body Dirac equation. These calculations, which have to be accurate relativistic Dirac-Fock calculations with no Slater approximation, have been performed by a number of groups[3]. This Dirac-Fock value results to about 99 % of the binding energy of the innermost electrons in heavy systems. The remaining 1 % of the observable effect arises from the QED corrections vacuum polarization and vacuum fluctuation as well as the part of the electron-electron interaction, which is not taken into account in the Dirac-Fock calculation, which is the magnetic interaction between the electrons and retardation. In addition to these four effects one has to take into account the influence of the extended nucleus with a realistic nuclear charge distribution directly in the Dirac-Fock calculations.

Magnetic interaction and retardation

According to the proposal of Gaunt[4] the unretarded interaction between two Dirac currents given by the Dirac matrices $\vec{\alpha}$ can be written like

$$H_G = - \frac{e^2}{r_{12}} \vec{\alpha}_1 \cdot \vec{\alpha}_2 \ . \tag{1}$$

Breit[5] proposed the quantum mechanical analogon to Darwin's retarded Hamilton function, which now usually is called the Breit operator

$$H_{Br} = -\frac{e^2}{2r_{12}} (\vec{\alpha}_1 \cdot \vec{\alpha}_2 + (\vec{\alpha}_1 \cdot \hat{n}) (\vec{\alpha}_2 \cdot \hat{n})) \text{ with } \hat{n} = \frac{\vec{r}_{12}}{|\vec{r}_{12}|} \ . \tag{2}$$

An even more accurate expression has been derived by Bethe and Salpeter[6], which is due to the exchange of a transverse photon $(R = r_{12})$

$$H'_{Br} = - \frac{1}{2} e^2 \vec{\alpha}_{1i} \cdot \vec{\alpha}_{2j} \left[\delta_{ij} \frac{\cos \omega R}{R} + \frac{\partial^2}{\partial R_i \partial R_j} \frac{\cos \omega R - 1}{\omega^2 R} \right], \tag{3}$$

where ω is the energy transferred by the virtual photon.

In direct two-electron matrix elements of H'_{Br} the photon energy $\omega = 0$; in this case H'_{Br} reduces exactly to H_{Br_2}. The same result is obtained when all contributions of the order ω^2 or higher $(O((\frac{v}{c})^4)$ are neglected. Therefore the normal Breit operator H_{Br} is a good approximation for small Z, because of $v \ll c$ in this region. An alternative expression for the Breit operator is

$$H''_{Br} = - \frac{e^2}{2r_{12}} (\vec{\alpha}_1 \cdot \vec{\alpha}_2 \cos \omega \, r_{12} + (1 - \cos \omega \, r_{12})) \qquad . \qquad (4)$$

Both expressions H'_{Br} and H''_{Br} are good for the region of large Z.

	$<H'_{Br}>$	$<H_{Br}>$
Ne ($Z = 10$)	0.033	0.033
Xe ($Z = 54$)	11.420	11.549
Pb ($Z = 82$)	48.393	49.521
No ($Z = 102$)	107.203	110.516

Tab. 1

Contribution of the magnetic and retardation contribution to the total energy of an atom in a.u. for the operators H_{Br} and H'_{Br}.

Table 1, which is taken from the paper of Mann[7] contains the expectation values of the two operators H'_{Br} and H_{Br} for different Z in a.u. Only for very large Z appreciable discrepancies occur. One has to have in mind that one s electron contributes to these values already by more than 40 %. In addition, one gets somewhat different numbers, when different wave functions are taken into the calculation.

Vacuum polarization

Within the last few years calculations of the vacuum polarization effect by Gyulassy[8] and Rinker[9] have been performed, which explicitly took into account Coulomb wavefunctions to describe the intermediate states of the virtual electron and positron cloud.

This method of calculation leads to values which are correct
even for the region of high Z elements. If one compares these
calculations with the usual $Z\alpha$ and α expansion, usually applied
for low Z calculations, one has to state that the lowest order
Uehling term plus higher orders in $(Z\alpha)^n$ with n = 2,3,... plus
all higher order terms in α^n are included.

Vacuum fluctuation (self-energy)

If Coulomb wavefunctions are taken explicitly as intermediate
states in the calculation of the lowest order vertex correction,
the results for the vacuum fluctuation correction are expected
to be quite accurate, even in the region of $Z\alpha \approx 1$.
Mohr[10] used this method to calculate the self-energy with analytical
Coulomb wavefunctions for very high Z systems. Desiderio and
Johnson[11] as well as Cheng and Johnson[12] even went beyond that
approximation. They took into account numerical Dirac-Fock-Slater
wavefunctions with an extended nucleus as intermediate states. This
is the only way to continue the calculations into the region
Z > 137. Usually the result is expressed as

$$\frac{\alpha}{\pi} \; \frac{(Z\alpha)^4}{n^3} \; mc^2 \; F(Z\alpha) \qquad .$$

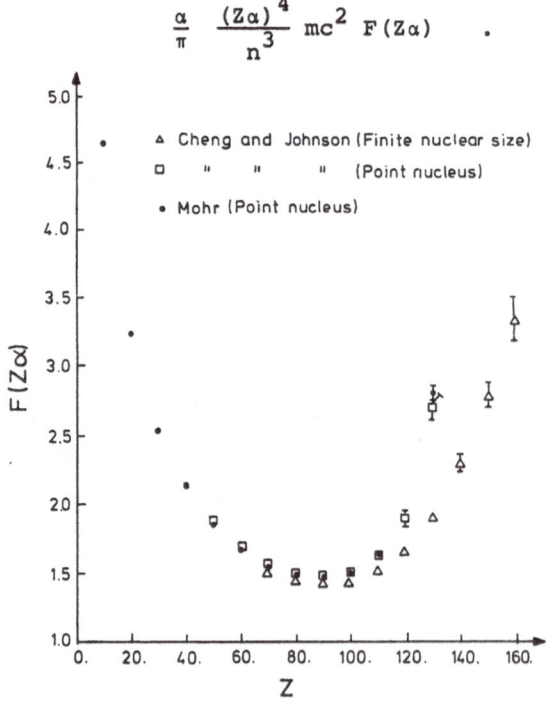

Fig. 1

Values for the
function $F(Z\alpha)$
in the expression
of the self-energy
for large and
very large Z (see
ref. 12 and 10).

A comparison of the function $F(Z\alpha)$ for the various calculations for the 1s electronic state can be seen in figure 1, where for low Z elements the results of Cheng et al.[12] and Mohr[10] agree well, whereas for high Z Mohr's values of $F(Z\alpha)$ increase much stronger than Cheng's results. The reason for this difference is the effect of the extended nucleus which is taken into account in the numerical wavefunctions of Cheng et al.[12]. Due to numerical uncertainties, the calculations of Cheng et al.[12] were not continued with the present version of the program above Z = 160. Therefore the very important question, if the self-energy of the innermost level may become so big for Z ≈ 173 that a diving of this level into the negative continuum can be prevented, cannot be answered up to now. There are experimental indications in the heavy ion collision of Cm on Pb which could be interpreted in this way. But actual calculations have not been performed so far.

Order of magnitude of the effects

In table 2 we list the contributions to the binding energies of the four effects discussed above for the innermost electrons of the elements Z = 90 and Z = 100.

		Z = 90	Z = 100
Magnetic contribution:	1s	+492 eV	+715 eV
	$2p_{1/2}$	+100 eV	+153 eV
Retardation:	1s	− 36 eV	− 41 eV
	$2p_{1/2}$	− 10 eV	− 13 eV
according to ref. 7,13,14			
Vacuum polarization:			
1. order Uehling term with extended nucleus	1s	− 80 eV	−148 eV
	$2p_{1/2}$	− 2 eV	− 4 eV
higher orders (ref.8 and ref. 9)	1s	+ 4 eV	+ 8 eV
Vacuum fluctuation:	1s	+306 eV	+457 eV
according to ref. 10 to ref. 12	$2p_{1/2}$	+ 7 eV	+ 15 eV

Table 2: Contributions of the four corrections to the 1s and $2p_{1/2}$ level of Z = 90 and 100.

These values have to be compared with the binding energy of about 141 keV for the 1s state of Z = 100, which is the result of the solution of the SCF Dirac-Fock equation[13]. The agreement between the experimental results and theory is good within a few eV.

Where do we stand now?

Recently, Deslattes et al.[15] compared all available results of experimental inner shell X-ray energies with theoretical calculations. They showed (see fig. 2) that there seems to be a linear trend proportional to Z for the difference between experimental and theoretical values for the K_{α_1} line.

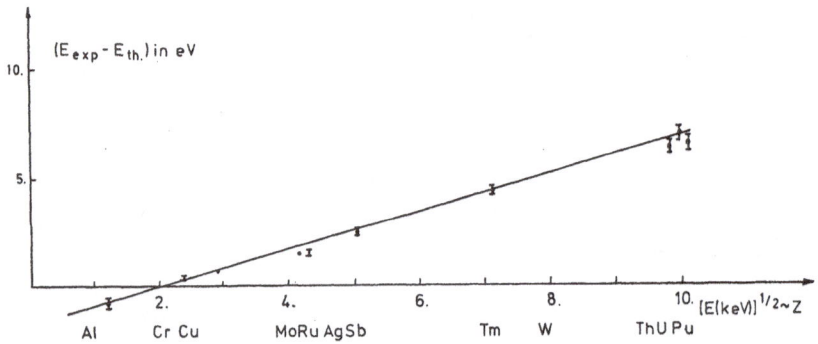

Fig. 2: Difference between experimental and theoretical values for the K_{α_1} line.

Up to now there is no answer for this discrepancy.

To avoid any calibration problem between different experiments which may be the reason for this systematic discrepancy, Borchert et al.[16] measured X-ray energies for low Z and high Z elements simultaneously in different orders of the Bragg reflection. Their results show an agreement for the measured X-ray energy difference and theory which is always better than 2 eV.

According to latest comparisons, the discrepancy shown in fig. 2
decreases again for the high Z elements. If this is true it is
easy to be understood[17] why Borchert et al.[16] did not measure any
big difference. They always compared one element on the increasing
low Z part of the curve with a high Z element on the decreasing
upper part of the curve, so that the relative difference between
both discrepancies remained smaller than 2 eV.

Finally one should mention the large discrepancy which shows up
between experimental $K_{\alpha_1}^h$ hypersatellite lines and theoretical
values[18,19]. For Hg it is still of the order of 30 eV.

To close the gap between experiment and theory in the future, to
my mind one main effort must be undertaken from the theoretical
side. Because we are dealing with complicated systems of many
electrons, which are connected in a self-consistent way, one has
to look into the self-consistency effects on the whole atom and its
total energy which will arise from all three effects, the magnetic
interaction, the vacuum polarization, and vacuum fluctuation. The
second is easier, because the main part of it can be inserted as
an additional local potential in the SCF calculations. Also the
first can be (and already has been) included in the SCF
calculation[20]. The most complicated will be the third. Up to now
there is no direct way to include the vacuum fluctuation in the
calculation itself. Although these indirect QED effects are small,
one has to study them in the light of these discrepancies with
great care.

How large are the contributions for Z $\approx$ 170?

The magnetic contribution and retardation never has been
calculated for the region of superheavy elements, but from an
extrapolation of ref. 7 one may expect that the contribution to
the 1s binding energy for Z = 170 is in the order of +20 to +40 keV.
The vacuum polarization contribution as calculated by ref. 8 and 9
is expected to be $\approx$ -10 keV. The vacuum fluctuation, as calculated
by Cheng and Johnson[12], can only be extrapolated for these very

high Z systems (see the discussion about this question in the part on vacuum fluctuation). If we assume $F(Z\alpha) \approx 4.5$ for $Z = 170$, we get a contribution of $\approx +18$ keV.

Thus, the total QED contribution to the 1s level of $Z = 170$ is expected to be in the order of $\approx +40$ keV. This number has to be compared with the influence of the extended nucleus. For $Z = 170$ an uncertainty in the nuclear radius of $\Delta R = 0.1$ fm [21] leads to a change in the 1s binding energy of 3 keV. Thus, an uncertainty of about 1 fm in the nuclear radius already amounts to the same order of magnitude as the sum of the QED contributions.

Consequences

We have seen that QED effects in many-electron atoms in the region between $Z = 80$ to 100 are in the order of 10^{-3} to 10^{-2} of the binding energy for the innermost electrons. Of course, it would be most interesting to measure one-electron systems, even at these high Z. Because this will be very complicated to achieve experimentally, one might spend further effort to get better results from one-hole systems instead. Of course, theoretically this is much more complicated.

Although for superheavy systems theoretical values are still very inaccurate and experiments are not available, it is still of great principal interest. Great effort should be undertaken to get also some results from this region. Maybe, experiments at the heavy ion accelerators one day will give some answers to this important question.

References

1. One example is F.T. Porter and M.S. Freedman, Phys. Rev. Lett. 27, 293 (1971)

2. G.L. Borchert, P.G. Hansen, B. Jonson, H.L. Ravn and J.P. Desclaux in Atomic Masses and Fundamental Constants 6 (Eds. J.A. Nolen and W.B. Benenson) N.Y. and London Plenum Press, p. 189

3. J.P. Desclaux, Comp. Phys. Comm. 9, 31 (1975), program developed by J.B. Mann, described in J.B. Mann and J.T. Waber, J. Chem. Phys. 53, 2397 (1970); in addition there exists a number of analogous programmes e.g. by Meyers

4. J.A. Gaunt, Proc. Roy, Soc. (London) A122, 513 (1929)

5. G. Breit, Phys. Rev. 34, 553 (1929); 36, 383 (1930); 39, 616 (1932)

6. H.A. Bethe and E.E. Salpeter in Quantum Mechanics of one and two electron atoms, Academic Press N.Y. 1957, p. 181

7. J.B. Mann and W.R. Johnson, Phys. Rev. A4, 41 (1971)

8. M. Gyulassy, Phys. Rev. Lett. 33, 921 (1974)

9. G.A. Rinker and L. Wilets, Phys. Rev. A12, 748 (1975)

10. P.J. Mohr, Ann. Phys. (N.Y.) 88, 26, and 52 (1974)

11. A.M. Desiderio and W.R. Johnson, Phys. Rev. A3, 1267 (1971)

12. K.T. Cheng and W.R. Johnson, Phys. Rev. A14, 1943 (1976)

13. B. Fricke, J.P. Desclaux and J.T. Waber, Phys. Rev. Lett. 28, 714 (1972)

14. K.N. Huang, M. Aoyagi, M.H. Chen, B. Crasemann and H. Mark, At. Data Nucl. Data Tables 18, 243 (1976) and references therein

15. R.D. Deslattes, E.G. Kessler, L. Jacobs and W. Schnitz, Phys. Lett. 71A, 411 (1979)

16. G.L. Borchert, P.G. Hansen, B. Jonson, H.L. Ravn and J.P. Desclaux, 6 Int. Conf. on Atomic Masses, 1979

17. R.D. Deslattes, private communication, Dec. 1980

18. K. Schreckenbach, H.G. Börner and J.P. Desclaux, Phys. Lett. <u>63A</u>, 330 (1977)

19. N. Beatham, I.P. Grant, B.J. Mc Kenzie and S.J. Rose, Physica Scripta <u>21</u>, 423 (1980)

20. W. Johnson, private communication
    J.P. Desclaux,   "   "   "

21. B. Fricke, W. Greiner and J.T. Waber, Theor. Chim. Acta <u>21</u>, 235 (1971)

POSITRONENERZEUGUNG BEI SCHWERIONENSTÖSSEN [+]

H. Backe

Institut für Kernphysik
Technische Hochschule Darmstadt

Abstract:

The positron production yields in close ion-atom collisions have been investigated for the U+Cm, U+U and U+Pb scattering systems at U beam energies between 4.7 and 5.9 MeV/u. Furtheron positron spectra were measured for the overcritical U+U and the undercritical U+Pb system at 5.9 MeV/u and a scattering angle $\theta_{Lab}^{Ion} = 45^{\circ} \pm 10^{\circ}$. All positron yields, including earlier measurements, follow a simple scaling law. No characteristic deviations from this or other signatures indicating level diving for the overcritical U+Cm and U+U systems have been observed within the experimental errors of about 30 %.

+) Auszug aus Darmstädter Habilitation D 17

1.   Einleitung

Kürzlich durchgeführte Experimente zur Positronenerzeugung (Bac H 78, Koz K 79) und die Messung der $1s_\sigma$ - Anregungswahrscheinlichkeiten (Lie A 80) bei nahen Schwerionenstößen mit vereinigten Kernladungen $Z_u = Z_1 + Z_2 > 137$ unterstützen im wesentlichen die Idee von adiabatisch gebildeten Quasi-Atomen. In idealen Quasiatomen haben die Elektronen genügend Zeit, sich während des Stoßvorganges in jedem Moment auf das Zweizentrenpotential der Stoßpartner einzustellen. Nach theoretischen Rechnungen (Sof R 78, Wie M 79) erwartet man im Moment der nächsten Annäherung der Atomkerne ein starkes Schrumpfen der Elektronenwellenfunktionen und Bindungsenergien der $1s_\sigma$-Elektronen, die in den schwersten Stoßsystemen den Wert $2m_o c^2 = 1.022$ MeV überschreiten können.

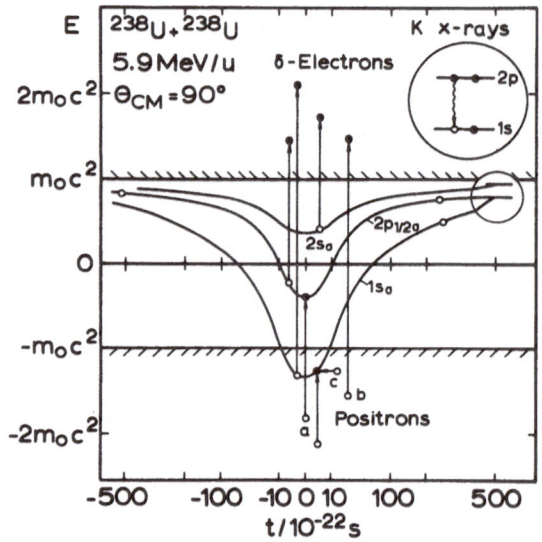

Fig. 1.1

Bindungsenergien der $1s_\sigma$-, $2p_{1/2\sigma}$- und $2s_\sigma$-Zustände als Funktion der Zeit für ein $^{238}U + ^{238}U$ Quasiatom. Infolge starker Störungen werden Elektronen innerer Schalen ins positive Kontinuum angeregt (δ-Elektronen). Die verbleibenden Löcher führen entweder zur Emission charakteristischer Röntgenstrahlen in den separierten Atomen oder werden mit Elektronen des negativen Energiekontinuums aufgefüllt, wobei Positronen entstehen: (a) induziert, (b) direkt, (c) spontan (aus Bac 78).

In Fig.1.1 sind die in einer adiabatischen Zweizentrenbasis berechneten Bindungsenergien der $1s_\sigma$-, $2s_\sigma$- und $2p_{1/2\sigma}$-Zustände als Funktion der Stoßzeit aufgetragen. Für das gewählte Beispiel übersteigt die Bindungsenergie für eine sehr kurze Zeit von $2 \cdot 10^{-21}$s den Wert $-2m_o c^2$. Eine derartige Situation wird überkritisch genannt.

Während in früheren Experimenten zur Positronenerzeugung (Bac H 78, Koz K 79) im wesentlichen unterkritische Stoßsysteme (kein Eintauchen des $1s_\sigma$-Niveaus) studiert wurden, zielen die hier zu beschreibenden neueren Experimente auf eine Untersuchung überkritischer Systeme. Ein durch einen Ionisationsprozeß erzeugtes Loch in der $1s_\sigma$-Schale kann man in einem überkritischen System wie ein gebundenes Positron auffassen, das, falls es freigesetzt wird, zu Abweichungen der energieintegrierten Positrinenerzeugungswahrscheinlichkeit P_{e+} von derjenigen unterkritischer Systeme führen sollte.

Für die unterkritischen Stoßsysteme wurde experimentell ein exponentieller Verlauf der Positronenerzeugungswahrscheinlichkeit als Funktion von relevanten kinematischen Variablen wie der Stoßzeit $2\hat{t}$ (Kan 78, Bac 78, Kan 79), dem Abstand minimaler Annäherung R_m (Bok 79) oder dem Stoßparameter b (Arm K 79) gefunden. Außerdem könnte man erwarten, daß in den Positronenspektren überkritischer Systeme bei niedrigen Positronenenergien charakteristische Abweichungen im Vergleich zu denen unterkritischer Systeme auftreten und das Eintauchen somit signalisiert wird. Ein derartiges Signal würde eine direkte Aussage über die Bindungsenergie der $1s_\sigma$-Elektronen in der Nähe des Abstandes minimaler Annäherung beim Stoß der Ionen enthalten. Eine solche experimentelle Information wäre aus dem Grunde von Bedeutung, weil Bindungsenergien $> 2m_o c^2$ mit anderen experimentellen Methoden bisher nicht nachgewiesen werden konnten (Lie A 80). Außerdem ist das Verständnis des Prozesses der Positronenerzeugung in überkritischen Systemen natürlich selbst von großem Interesse (Rei G 77, Raf F 78).

Aus diesen Gründen wurde P_e+ für die überkritischen Systeme U+Cm ($Z_u=188$), U+U ($Z_u=184$) und das unterkritische System U+Pb ($Z_u=174$) unter verschiedenen dynamischen Bedingungen untersucht. Weiterhin wurde ein Positronenspektrum für das überkritische U+U Stoßsystem gemessen und mit demjenigen des unterkritischen U+Pb Stoßsystems verglichen (Bac B 80).

In Abschnitt 2 dieser Arbeit werden die physikalischen Meßgrößen definiert, in Abschnitt 3 wird die experimentelle Anordnung beschrieben und in Abschnitt 4 auf die anzubringenden Korrekturen eingegangen. Die entscheidende Schwierigkeit bei diesen Experimenten liegt darin, daß über Coulombanregungsprozesse Kernniveaus der Stoßpartner angeregt werden können, die mit Übergangsenergie größer $2m_oc^2$ zerfallen. Dabei ist innerer Paarzerfall möglich, wobei die entstehenden Positronen ein ernsthaftes Untergrundproblem darstellen. Auf die diesbezüglichen Korrekturen wird ausführlich eingegangen. In Abschnitt 5 folgt die Zusammenstellung und Diskussion der Ergebnisse, und in Abschnitt 6 werden die Perspektiven der Positronenspektroskopie abgehandelt.

2.   Einiges zu den Meßgrößen

Bei der Positronenspektroskopie geht es darum, die energiedifferentielle Positronenerzeugungswahrscheinlichkeit ("Positronenspektren")

$$\Delta P_e+/\Delta E_e+(E_e+) = (\Delta Z_e+/E_e+)/Z_{part} \qquad (2.1)$$

und die energieintegrierte Positronenerzeugungswahrscheinlichkeit

$$P_e+ = Z_e+/Z_{part} \qquad (2.2)$$

beim Schwerionenstoß zu messen. Hierbei ist ΔZ_e+ die Anzahl der Positronen im Energieintervall ΔE_e+ bei der Energie E_e+, Z_e+ die über das Positronenspektrum integrierte Anzahl atomarer Positronen und Z_{part} die in einem Winkelbereich $\theta_{Lab}^{Ion} \pm \theta_{Lab}^{Ion}$ gestreuten Teilchen, zu denen die Positronen in Koinzidenz gemessen werden. Diese Größen werden z.B. als Funktion des minimalen Abstandes R_m bei Rutherfordstreuung untersucht mit:

$$R_m = a (1+\epsilon) , \qquad (2.3)$$

wobei

$$2a = [(M_1+M_2)/(M_1M_2)] \cdot [Z_1Z_2e^2/(E_1/M_1)] \qquad (2.4)$$

der minimale Abstand bei einem zentralen Stoß und

$$\epsilon = 1/\sin(\theta_{CM}/2) \qquad (2.5)$$

die Exzentrizität beim Streuwinkel θ_{CM} im Schwerpunktsystem sind. Weiterhin bedeuten M_1, Z_1 sowie M_2, Z_2 Massen- und Ordnungszahl von Projektil- bzw. Targetkern und E_1 die Projektilenergie. Für symmetrische Systeme ist R_m nur bei einem Streuwinkel von $\theta_{Lab}^{Ion} = 45^o$ eindeutig definiert. Aus diesem Grunde wurden fast alle hier beschriebenen Messungen unter diesem Streuwinkel mit $\theta_{Lab}^{Ion} = \pm 10^o$ ausgeführt.

Für das Verständnis des Positronenerzeugungsprozesses spielt die aus den beobachteten kinematischen Variablen abgeleitete Stoßzeit ($v = (2E_1/M_1)^{1/2}$ ist die Projektilgeschwindigkeit im Unendlichen)

$$2\dot{t} = (2a/v)(\epsilon + 1.6 + 0.45/\epsilon) \qquad (2.6)$$

eine wesentliche Rolle (Kan 79). Diese Größe ist über die Extrema der Funktion $\dot{R}(t)/R(t)$ definiert, wobei $R(t)$ der Abstand von Projektil- und Targetkern und $\dot{R}(t)$ die radiale relative Geschwindigkeit zum Zeitpunkt t sind. Die Positronenerzeugung wird auch als Funktion des Stoßparameters

$$b = a \, ctg \, (\theta_{CM}/2) \qquad (2.7)$$

diskutiert (Arm K 79).

Um eine Vorstellung von der Größenordnung der oben definierten physikalischen Größen zu vermitteln, seien für ein typisches U+U Experiment bei einer Laborenergie des U-Projektiles von 5.9 MeV/u und einem Streuwinkel von $\theta_{Lab}^{Ion} = 45^o$ ihre Zahlenwerte angegeben: $2a = 17.36$ fm, $\epsilon = 1.41$, $R_m = 20.95$ fm, $2\hat{t} = 1.71 \cdot 10^{-21}$ s, $b = 8.68$ fm. Der Rutherfordwirkungsquerschnitt für das Projektil im Laborsystem ist $d\sigma/d\Omega = 2.13$ b/sr. Die Positronenerzeugungswahrscheinlichkeit $P_{e^+} \simeq 2 \cdot 10^{-4}$.

3. Die experimentelle Anordnung

3.1 Das Solenoid - Spektrometer

Das am Strahl des Darmstädter Schwerionenbeschleunigers UNILAC aufgebaute Solenoid-Positronen (Elektronen)-Transportsystem ist in Fig.3.1 dargestellt. Das Magnetfeld wird mit mehreren normalleitenden Spulen erzeugt. Im Target bei $\gamma = 13$ cm entstandene Positronen oder Elektronen werden in ihm zunächst auf Spiralbahnen in einen untergrundfreien Raum transportiert, wo sich die Detektoranordnung zu

ihrem Nachweis befindet. Die Spulen werden von zwei 1500 A Stromversorgungen gespeist. Die erste versorgt die drei Spulen zwischen
0 cm $\leq \mathcal{Y} \leq$ 62 cm, wobei die beiden langen Spulen paralell oder in Serie
geschaltet werden können. Die zweite Stromversorgung speist unabhängig
davon die restlichen Spulen. Das in Fig. 3.1 eingezeichnete Magnetfeld
wurde für Parallelschaltung der beiden langen Spulen und I = 1500 A
berechnet. Die Magnetfeldüberhöhung bei $\mathcal{Y} \leq$ 12 cm bewirkt die Spiegelung der im Target gebildeten und in falsche Richtung emittierten Positronen bzw. Elektronen, wenn ihr Emissionswinkel gegen die Magnetfeldachse der Bedingung

$$\theta > \theta_{sp} = arc\ sin\ (B_T/B_{max}) \tag{3.1}$$

genügt, wobei B_T das Magnetfeld am Target und B_{max} das maximale
Feld sind. In dem in Fig. 3.1 dargestellten Magnetfeld ist θ_{sp} = 58.7°,
so daß 74% aller im Target gebildeten Positronen (Elektronen) letztlich
in Richtung Detektoranordnung laufen.

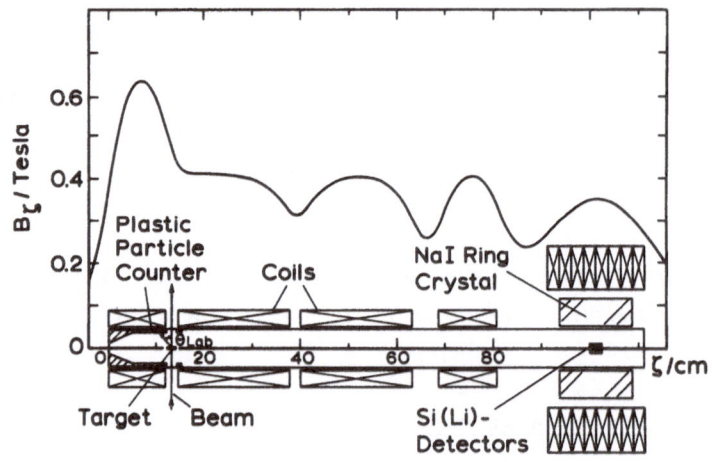

Fig. 3.1
Das Solenoid-Spektrometer zum Nachweis von Positronen und Elektronen.
Mit eingezeichnet ist der Magnetfeldverlauf auf der Achse des Solenoiden. Der Strahl tritt senkrecht zur Solenoidachse bei $\mathcal{Y}$ = 13 cm in die
Vakuumkammer ein (aus Bac B 80).

3.2. Der Positronendetektor

Beim Durchgang geladener Projektile durch die Targetmaterie werden δ-Elektronen ausgelöst, die im Magnetfeld ebenfalls auf den Detektor transportiert werden. Dieser δ-Elektronenuntergrund ist ca. einen Faktor 10^4 intensiver als die zu messenden Positronen. Die Selektion der Positronen erfolgt deshalb über den Nachweis der 511 keV Vernichtungsstrahlung, die nach dem Abbremsen von Positronen in Metallen mit sehr hoher Wahrscheinlichkeit emittiert wird.

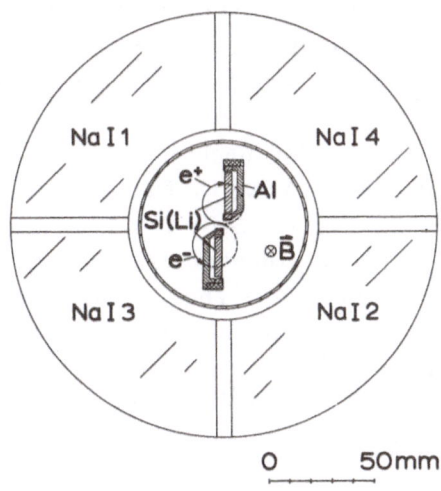

Fig. 3.2
Die NaJ-Si(Li)-Detektoranordnung zur Spektroskopie von Positronen nach Schwerionenstößen im Querschnitt senkrecht zur Solenoidachse bei $\mathcal{Y} = 102$ cm (vgl. Fig. 3.1). Die kreisrunden Si(Li)-Detektoren haben einen Durchmesser von 20 mm und eine Dicke von 3 mm. Der vierfach segmentierte NaJ-Ringkristall hat folgende Dimensionen: innerer Durchmesser 90 mm, äußerer Durchmeseer 204 mm, Länge 150 mm (aus Bac B 80).

Die Detektoranordnung zur Messung von Positronenspektren ist in Fig. 3.2 dargestellt. Sie besteht aus zwei Si(Li)-Dioden und einem vierfach segmentierten NaJ-Ringkristall, der die Detektoren umgibt, zum Nachweis der 511 keV Vernichtungsstrahlung. Die kreisrunden Si(Li)-Dioden sind mit ihren Flächennormalen senkrecht zur Magnetfeldachse angebracht. Beide Detektoren sind von der Achse weg verschoben, der eine nach oben und der andere nach unten. Die empfindliche Fläche des oberen Zählers sieht nach links, diejenige des unteren Zählers nach

rechts. Es ist nun leicht einzusehen, daß Positronen mit einer Rechts-
spirale eine gute Chance haben, die empfindliche Fläche des Zählers zu
treffen, während Elektronen, die mit einer Linksspirale sich dem Detektor nähern, auf seine 3 mm dicke Aluminium-Rückseite auftreffen. Nur
ein kleiner Bruchteil der δ-Elektronen mit einer Energie größer als ca.
1.4 MeV kann das Aluminium durchdringen und ein Signal im Zähler erzeugen. Ein größeres Problem bereiten im Target gestreute Elektronen, die
nicht mehr vom Magnetfeld der ersten Spule zurückgespiegelt werden. Sie
können direkt auf die empfindliche Fläche des Detektors auftreffen. Die
Wahrscheinlichkeit eines solchen Prozesses kann mit den 365 keV Konver-
sionelektronen einer ^{113}Sn-Quelle gemessen werden. Die ^{113}Sn-Präparate-
lösung wurde dazu auf eine 1 mg/cm^2 dicke Ni-Folie nahezu punktförmig
aufgebracht, und die Folie auf einen Targetrahmen montiert, um ähnliche
Verhältnisse wie bei Strahlbetrieb zu simulieren. Die gemessenen Spektren mit den beiden Polungen des Magnetfeldes sind in Fig. 3.3 dargestellt. Die Nachweiswahrscheinlichkeit für gestreute Elektronen betrug
ca. 0.7 %. Es ist damit klar, daß ein Positron zusätzlich durch seine
511 keV Vernichtungsstrahlung identifiziert werden muß. Der Unter-
drückungsfaktor für Elektronen war aber groß genug, um Elektron-Posi-
tron Summenkoinzidenzen bei der Auswertung der Experimente vernachläs-
sigen zu können. In einem U+U Experiment bei einer Einschußenergie von
5.9 MeV/u und einem Streuwinkel $\theta_{Lab}^{Ion} = 45^{o} \pm 10^{o}$ trugen Summenkoinzidenzen zu weniger als 2 % bei.

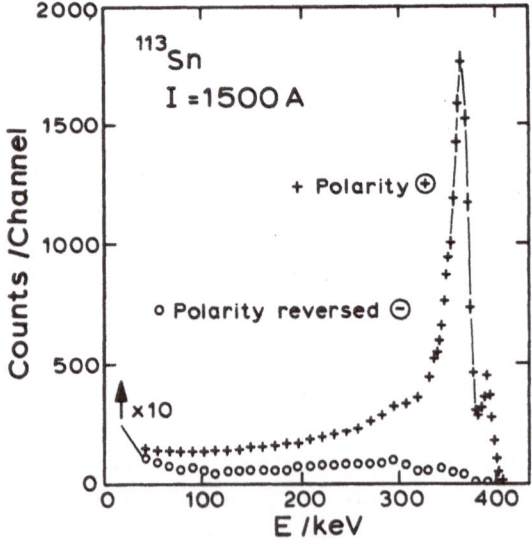

Fig. 3.3

Einfluß der Feldumpolung auf die Nachweiswahrscheinlichkeit der Si(Li)-
Detektoren in Fig. 3.2 für 365 keV Elektronen. Die untere Kurve
entspricht der Polung bei der Positronenspektroskopie.

Die Positronennachweiswahrscheinlichkeit wurde mit intensitätsge-
eichten ^{22}Na- und ^{68}Ge/Ga-Quellen gemessen. Sie ist in Fig. 3.4 als
Funktion der Positronenenergie dargestellt.

Bei der Messung wurde in einem Segment des NaJ-Ringzählers ein
Signal in der 511 KeV Linie verlangt, während im gegenüberliegenden
Kristall ein Signal im gesamten Einergiebereich zugelassen war. Dadurch
vergrößert sich gegenüber der (scharfen) 511 keV Bedingung in beiden
Kristallen die Nachweiswahrscheinlichkeit um fast einen Faktor 2. Eine
Verfälschung des Positronenspektrums infolge von Summenkoinzidenzen mit
comptongestreuten Vernichtungsquanten im Si(Li)-Detektor wurde nicht
beobachtet.

Zur Messung der energieintegrierten Positronenerzeugungswahrscheinlichkeit P_e+ kann natürlich über das gemessene Energiespektrum
$\frac{\Delta P_e+}{\Delta E_e+}$ integriert werden. Wesentlich effizienter ist es jedoch, die gesamte Si(Li)-Detektoranordnung nur als einen Stopper für Positronen zu
verwenden. Man erhält dann nach einem von Heßberger (Heß 77) beschriebenen Verfahren die in Fig. 3 .4 (obere Kurve) dargestellte Nachweiswahrscheinlichkeit für Positronen, die ungefähr einen Faktor 3 über
der energiedifferentiellen unteren Kurve liegt.

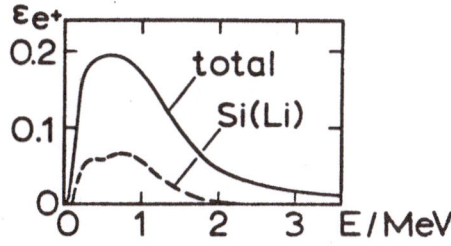

Fig. 3.4
Nachweiswahrscheinlichkeit der Detektoranordnung Fig. 3.2 für Positronen. Die untere gestrichelte Kurve gilt für energieanalysierte
Positronen mit dem Si(Li)-Dektor. Die obere durchgezogene Kurve zeigt
die totale Nachweiswahrscheinlichkeit für Positronen, bei der die
gesamte Si(Li)-Detektoranordnung nur als passiver Fänger für Positronen
fungiert (aus Bac B 80).

3.3. Die Teilchenzähler

Die Positronen werden in Koinzidenz zu gestreuten Teilchen nachgewiesen. Dazu wird der in Fig. 3.5 dargestellte Plastik-Szintillationszähler für in Vorwärtsrichtung gestreute Teilchen verwendet. Nach Kernkontakt oder Coulombspaltung bei fast zentralem Stoß in Rückwärtsrichtung emittierte Spaltfragmente können mit zwei großflächigen Silizium-Oberflächensperrschichtzählern nachgewiesen werden.

Auf die umfangreiche Elektronik, die Datenaufnahme und die Datenanalyse wird hier nicht eingegangen. Im folgenden seien aber die wesentlichen Korrekturen zur Bestimmung des Anteiles atomarer Positronen diskutiert. Die Hauptkorrektur rührt dabei von den nuklearen Positronen her.

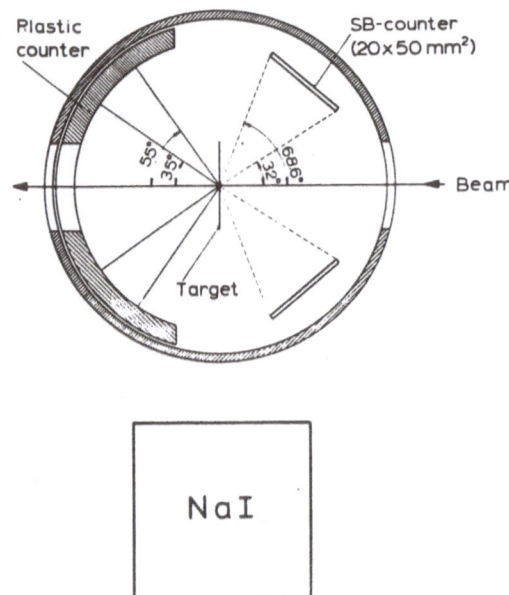

Fig.3.5

Die Teilchenzähler in einem Schnitt senkrecht zur Solenoidachse durch die Targetregion bei $\mathcal{Y}$ = 13 cm von Fig. 3.1. Elastisch gestreute Teilchen erzeugen in einer 50 μ dicken Plastik-Szintillatorfolie einen Lichtblitz, der mit einem Photomultiplier nachgewiesen wird. In Rückwärtsrichtung fliegende Spaltfragmente werden mit zwei Halbleiterzählern gemessen, die mit ca. 2 mg/cm^2 dicken Alu-Folien zur Absorption der Targetröntgenstrahlung abgedeckt sind. Mit dem 7.5 x 7.5 cm NaJ-Detektor wird die Target-γ-Strahlung in Koinzidenz zu Ereignissen in diesen Zählern gemessen.

4. Korrekturen

Die zur Berechnung von $\Delta P_{e^+}/\Delta E_{e^+}$ bzw. ΔP_{e^+} nach Gleichungen (2.1) und (2.2) notwendigen Positronenzahlen ΔZ_{e^+} und Z_{e^+} werden aus den atomaren Positronenzahlen $\Delta Z_{e^+}^{atom}$ und $Z_{e^+}^{atom}$ (siehe 4.3) wie folgt bestimmt:

$$Z_{e^+}^{atom} = \Delta Z_{e^+}^{atom}/(\varepsilon_{e^+}(E_{e^+})\, f_{tot}) \tag{4.1}$$

und

$$Z_{e^+} = Z_{e^+}^{atom}/(\overline{\varepsilon_{e^+}}\, f_{tot}) \tag{4.2}$$

Dabei sind $\varepsilon_{e^+}(E_{e^+})$ die Nachweiswahrscheinlichkeit für Positronen bei der Energie E_{e^+} (siehe Fig. 3.4 untere gestrichelte Kurve) und $\overline{\varepsilon_{e^+}}$ die mit dem Positronenspektrum aus der oberen Kurve von Fig. 3.4 gemittelte Nachweiswahrscheinlichkeit, für die $\overline{\varepsilon_{e^+}} = 0.167\pm0.021$ verwendet wurde. Der Faktor f_{tot} ist ein Totzeitkorrekturfaktor von der Größenordnung $0.8 < f_{tot} < 1$, der durch die Datenaufnahmeanlage bedingt ist. Es sei noch bemerkt, daß alle mit den Positronenspektren in Zusammenhang stehenden Größen (also z.B. $\Delta Z_{e^+}^{atom}$) als bereits von der Apparatefunktion des Si(Li)-Detektors entfaltet zu betrachten sind.

Zur Berechnung der atomaren Positronenzahlen $Z_{e^+}^{atom}$ und $\Delta Z_{e^+}^{atom}$ aus den primär gemessenen Zahlen $Z_{e^+}^{exp}$ und $\Delta Z_{e^+}^{exp}$ müssen eine Reihe von Korrekturen angebracht werden (im folgenden ist die Gleichung nur für Z_{e^+} hingeschrieben; eine analoge Beziehung gilt für ΔZ_{e^+}):

$$\begin{aligned}
Z_{e^+}^{atom} = &\; Z_{e^+}^{exp} - Z_{e^+}^{int}(Z_1,M_1) - Z_{e^+}^{int}(Z_2,M_2) \\
&- Z_{e^+}^{ext}(T) - Z_{e^+}^{ext}(S)
\end{aligned} \tag{4.3}$$

Die Messung muß also auf vier Anteile korrigiert werden: Auf die Anteile $Z_{e^+}^{int}(Z_1,M_1)$ und $Z_{e^+}^{int}(Z_2,M_2)$ infolge innerem Paarzerfall von angeregten Kernniveaus im Projektil- und Targetkern, auf den Anteil von Positronen $Z_{e^+}^{ext}(T)$ durch äußere Paarkonversion der Target-γ-Strahlung im Target und einem entsprechenden Anteil im Solenoiden $Z_{e^+}^{ext}(S)$. Wie sich durch Testmessungen und Rechnungen herausstellte, ist letzterer Anteil vernachlässigbar (Heß 77, Bac B 75)

$$Z_{e^+}^{ext}(S) \approx 0. \tag{4.4}$$

Die äußere Paarkonversion im Target liefert demgegenüber bei Verwendung von 1 mg/cm^2 dicken Targets Korrekturen von bis zu 10 %, die berücksichtigt wurden. Im folgenden wird der einfacheren Schreibweise halber

$$Z_{e^+}^{exp} = Z_{e^+}^{exp} - Z_{e^+}^{ext}(T) \tag{4.5}$$

gesetzt.

Zur Bestimmung der Korrekturen infolge innerem Paarzerfall werden die Target-γ-Spektren mit einem 7.5 x 7.5 cm NaJ-Zähler (siehe Fig. 3.5) unter exakt denselben Koinzidenzbedingungen bezüglich gestreuter Teilchen wie die Positronen gemessen. Ein typisches Spektrum ist in Fig. 4.1 dargestellt.

Nach der Entfaltung von der Nachweisfunktion und Korrektur auf die Nachweiswahrscheinlichkeit des NaJ-Detektors erhält man die energiedifferentielle γ-Verteilung $dZ_\gamma/dE_\gamma(E_\gamma)$, aus der mit Hilfe von theoretischen Paarzerfallskoeffizienten (Sch S 78) $d\beta_{M\lambda}/dE_{e^+}(E_\gamma,E_{e^+},Z)$ ($M\lambda$ = Multipolarität) das zugehörige Positronenspektrum berechnet werden kann:

$$\frac{dZ_{e^+}}{dE_{e^+}}(E_{e^+},M\lambda) = \int_{E_\gamma=2m_0c}^{\infty} dE_\gamma \frac{dZ_\gamma}{dE_\gamma}(E_\gamma) \frac{d\beta_{M\lambda}}{dE_{e^+}}(E_\gamma,E_{e^+},Z) \tag{4.6}$$

und

$$dP_{e^+}/dE_{e^+} = (dZ_{e^+}/dE_{e^+}(E_{e^+},M\lambda))/Z_{part} \tag{4.7}$$

Für das in Fig.4.1 dargestellte γ-Spektrum wurden zwei Positronenspektren unter Annahme von E1 und E2 Multipolarität berechnet und sind im unteren Teil des Bildes gezeigt. Sie unterscheiden sich fast um einen Faktor 2. Hieraus ist bereits deutlich die Problematik der Korrektur auf nukleare Positronen ersichtlich, da die Multipolarität des γ-Spektrums nicht bekannt ist.

Zur Bestimmung von P_{e^+} muß über die Positronenenergie unter Berücksichtigung der energieabhängigen Nachweiswahrscheinlichkeit $\varepsilon_{e^+}(E_+)$ integriert werden und wir erhalten die erwartete Gesamtzahl an nuklearen Positronen

$$Z_{e^+\gamma}^{calc}(M\lambda) = \int_{E_{e^+}=0}^{\infty} dE_{e^+}\ \varepsilon_{e^+}(E_{e^+}) \frac{dZ_{e^+}}{dE_{e^+}}(E_{e^+},M\lambda) \tag{4.8}$$

Die Größe wird mit der wirklich gemessenen Anzahl von Positronen $Z_{e^+}^{exp}$ über das Verhältnis $Z_{e^+}^{exp}/Z_{e^+\gamma}^{calc}(M\lambda)$ verglichen. Es ist unter der Annahme von E1-Multipolarität für Stoßsysteme, bei denen rein nukle-

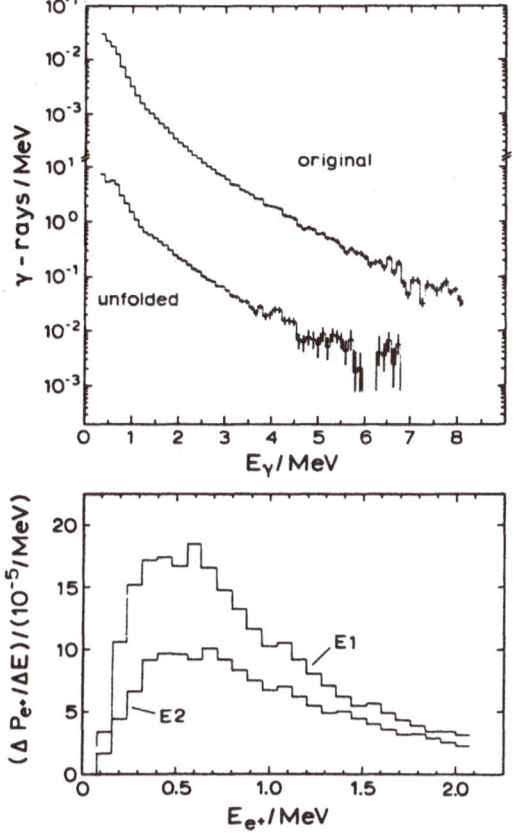

Fig. 4.1

Target γ-Spektrum, das mit dem 7.5 x 7.5 cm NaJ-Detektor aus Fig.3.5 beim Beschluß von ^{108}Pd mit 5.9 MeV/u ^{238}U aufgenommen wurde. Die Messung wurde in Koinzidenz zu den unter $45^O \pm 10^O$ gestreuten ^{108}Pd Rückstoßkernen durchgeführt. Das mit "unfolded" bezeichnete Spektrum entsteht aus dem "original" Spektrum nach Entfaltung von der Detektor-Apparatefunktion und Korrektur auf Detektoransprechwahrscheinlichkeit (Wei 80). Der untere Teil des Bildes zeigt die aus der γ-Verteilung mit Hilfe theoretischer Paarzerfallkoeffizienten berechneten Positronen-spektren unter der Annahme von E1 und E2 Multipolarität.

are Positronenerzeugung erwartet wird, nahezu 1. Diese Annahme ist damit aber keineswegs als gesichert anzusehen, denn eine Multipolari-tätsmischung von E0, M1, E2, E3 mit einem mittleren Verhalten wie E1-Multipolarität kann nicht ausgeschlossen werden. In Fig. 4.2 wurde durch Einführung eines Faktors f das Verhältnis $Z_{e^+}^{exp}/Z_{e^+\gamma}^{calc}$(E1) im Bereich nuklearer Positronenemmission unterhalb $Z_u=160$ auf 1 normiert. Das dargestellte Verhältnis übersteigt für die U+Pb, U+U und U+Cm Stoßsysteme deutlich den Wert 1. Dieser Überschuß-Anteil wird als atomare Positronenerzeugung interpretiert:

$$Z_{e^+}^{atom}/(Z_{e^+\gamma}^{calc}(E1)\cdot f) = Z_{e^+}^{exp}/(Z_{e^+\gamma}^{calc}(E1)\cdot f)-1 \ , \qquad (4.9)$$

woraus unmittelbar folgt:

$$Z_{e^+}^{atom} = Z_{e^+}^{exp} - Z_{e^+\gamma}^{calc}(E1)\cdot f. \qquad (4.10)$$

Durch Vergleich mit Gl. (4.3) unter Berücksichtigung von (4.4) und (4.5) erhält man, daß

$$Z_{e^+\gamma}^{calc}(E1)\cdot f = Z_{e^+}^{int}(Z_1,M_1) + Z_{e^+}^{int}(Z_2,M_2) \qquad (4.11)$$

ist. Es sei noch hinzugefügt, daß dieses Verfahren nur unter der Annahme richtig ist, daß sich die Multipolarität beim Übergang von $Z_u<160$ nach $Z_u>160$ nicht ändert, was keinesfalls sicher ist. Zur Diskussion der Fehler sei noch einmal die vollständige Formel für P_{e^+} hingeschrieben:

$$P_{e^+} = (Z_{e^+}^{exp} - Z_{e^+}^{ext}(T) - Z_{e^+\gamma}^{calc}(E1)\cdot f)/(Z_{part}\cdot f_{tot}\cdot\overline{\epsilon_{e^+}}). \qquad (4.12)$$

Für $Z_{e^+}^{exp}$ wird ein rein statistischer Fehler angenommen. Die Korrektur auf Positronen infolge externer Paarbildung $Z_{e^+}^{ext}(T)$ im Target ist nur auf 40 % genau, was sich wegen ihrer Kleinheit (< 10%) aber nicht stark auf den Gesamtfehler auswirkt. Für die Korrektur auf nuklearen Untergrund wurde $(\Delta Z_{e^+\gamma}^{calc}(E1)\cdot f)/(Z_{e^+\gamma}^{calc}(E1)\cdot f) = 0.15$ angesetzt. Weiterhin gilt für den Fehler der Teilchenzählrate $\Delta Z_{part}/Z_{part} = 0.07$ und der Nachweiswahrscheinlichkeit $\overline{\Delta\epsilon_{e^+}}/\overline{\epsilon_{e^+}} = 0.13$ % . Die Fehlerrechnung erfolgt nach dem Gaußschen Fehlerfortpflanzungsgesetz. Bei der Berechnung der Fehler der Positronenspektren wird analog verfahren. Im Gegensatz zu P_{e^+} ist aber bei $\Delta P_{e^+}/\Delta E_{e^+}$ der statistische Fehler überwiegend.

Die hier beschriebene Fehlerrechnung gilt nur für den relativen Vergleich der Messungen. Der Absolutwert ist mit einer zusätzlichen systematischen Unsicherheit von mindestens 20 % behaftet, die insbesondere durch die Korrektur auf nukleare Positronen bedingt ist.

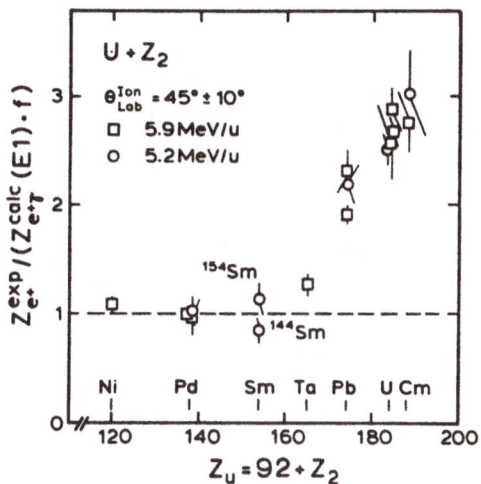

Fig. 4.2

Verhältnis gemessener zu aus γ-Spektren berechneter Positronenzahlen $Z_{e^+}^{exp}/(Z_{e^+\gamma}^{calc}(E1)\cdot f)$ als Funktion der vereinigten Kernladung Z_u = 92+Z_2. Bei Annahme von E1-Multipolarität ist f≈1. Für Z_u≳174 wird eine große Zahl von Positronen beobachtet, die nicht mit Kernanregungsprozessen erklärt werden können. (aus Bac B 80).

5. Ergebnisse und Diskussion

Alle bisher gemessenen energieintegrierten Positronenerzeugungs-wahrscheinlichkeiten sind in Fig. 5.1 als Funktion der Stoßzeit 2t̂ (vgl. Gleichung 2.6) dargestellt. Die erstmals gemessenen Positronen-spektren eines überkritischen U+U, eines unterkritischen U+Pb und eines nuklearen U+Pd-Systems sind in Fig. 5.2 dargestellt.

Es sollen zunächst die energieintegrierten Positronenerzeugungs-wahrscheinlichkeiten P_{e^+} diskutiert werden. Entsprechend theoretischen Berechnungen taucht das $1s_\sigma$-Niveau für die U+Cm Meßpunkte und für den U+U Meßpunkt bei niedrigster Stoßzeit in das negative Energiekontinuum ein. Alle anderen Punkte gehören zu unterkritischen Systemen. Zur Extrapolation von P_{e^+} von unterkritische in überkritische Systeme soll-te die analytische Darstellung von P_{e^+} bekannt sein, was natürlich zu komplizierten Rechnungen führt. Von Migdal (Mig 77) wurde aber gezeigt

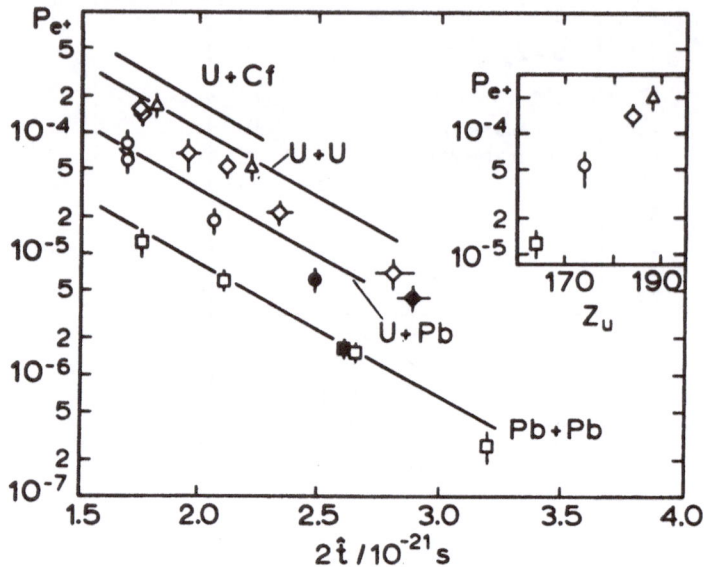

<u>Fig. 5.1</u>

Energieintegrierte Positronenerzeugungswahrscheinlichkeit P_e+ für verschiedene Stoßsysteme als Funktion der Stoßzeit $2\hat{t}$. Es ist Δ U+Cm,
◇ U+U, ◻ U+Pb, ◘ Pb+Pb . Offene Punkte repräsentieren θ_{Lab}^{Ion} = 45°±10°,
schwarze Punkte 25.5°±4.5° . Die ausgezognen Kurven sind theoretische
Berechnungen (Rei B 80). Das Zusatzbild zeigt die Z_u-Abhängigkeit von
P_e+ für $2\hat{t}$ = 1.75·10^{-21}s (aus Bac B 80).

(siehe auch Bac 78), daß das Übergangsmatrixelement in adiabatischer
Störungsrechnung 1. Ordnung

$$a_f(t=\infty) \quad = \quad - \int_{-\infty}^{+\infty} dt' \; <f|\tfrac{\partial}{\partial t}|i> \; \exp\left(i(E_f-E_i)t'/\hbar\right) \qquad (5.1)$$

in der Anregungswahrscheinlichkeit $|a_f(t=\infty)|^2$ einen Faktor

$$\exp[-(E_f-E_i)\tau/t] \qquad (5.2)$$

enthält, falls

$$(E_f-E_i)\tau/(2\hbar) \gg 1 \qquad (5.3)$$

gilt und das Übergangsmatrixelement $<f|\tfrac{\partial}{\partial t}|i>$ keine Singularitäten

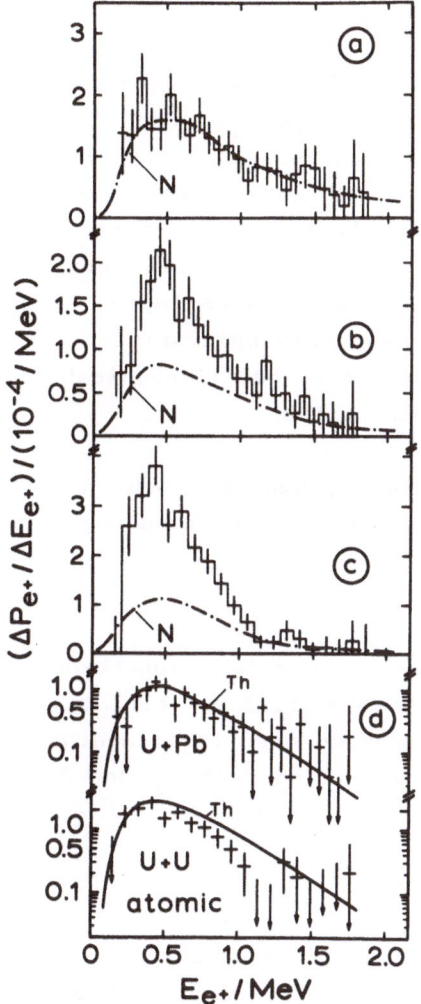

Fig. 5.2

Positronenspektren $\Delta P_{e^+}/\Delta E_{e^+}$ für (a) das nukleare U+Pd, (b) das unterkritische U+Pb und (c) das überkritische U+U Stoßsystem bei einer Uran-
strahlenergie von 5.9 MeV/u und $\theta_{Lab}^{Ion} = 45°\pm10°$. Die strichpunktierte
Kurven N zeigen den Anteil nuklearer Positronen, der aus den γ-Spektren unter Annahme von E1-Multipolarität berechnet wurde. Teil (d) des
Bildes zeigt in einer halblogarithmischen Darstellung den Anteil atomarer Positronen zusammen mit theoretischen Berechnungen (Th) (Rei B 80).
(Aus Bac B 80).

auf der reellen Achse aufweist. Es sind E_i, E_f die Energie der Anfangs-
bzw. Endzustände und τ eine für den Prozeß charakteristische Zeit.
Aus dieser einfachen Überlegung kann schon gesehen werden, daß nahezu
exponentielle Verläufe in den Spektren $\Delta P_{e^+}/\Delta E_{e^+}$ und P_{e^+} mehr oder
weniger triviale Aspekte der Positronenerzeugung beschreiben. Erst die
Absolutwerte der Positronenerzeugungswahrscheinlichkeiten und Abweichungen vom exponentiellen Verlauf beihalten detailliertere physikali-
sche Aussagen. Dabei muß man noch sicherstellen, daß derartige Abweichungen nicht eine Folge höherer Ordnungsprozesse in der Störungsrechnung sind, wenn der Vorgang nicht mehr rein adiabatisch verläuft. Für
das folgende sehen wir also, daß aus diesem Modell abgeleitete Aussagen
sicher nicht überinterpretiert werden dürfen.

Es kann nun gezeigt werden (Kan 78, Kan 79), daß in der Monopol-
näherung der Vorfaktor von (5.2) für Positronenerzeugung proportional

$$f(E_i)/(E_f-E_i) \qquad\qquad (5.4)$$

ist, wobei $1/(E_f-E_i)$ durch Anwendung der Hellmann-Feynman Relation
bei der Umformung des Matrixelementes in (5.1) auftritt und $f(E_i)$ eine
Art Fermifunktion ist, die eine Absenkung der Positronenintensität bei
kleinen Energien bedingt. Nach Intergration über E_i und E_f gilt dann
genähert

$$P_{e^+} \propto \exp(-1.4 \cdot 2\hat{t}\,\Delta E/\hbar) \qquad\qquad (5.5)$$

mit $2\hat{t}$ entsprechend (2.6) und ΔE der mittleren Energielücke für Paarbildung. Jedem Stoßsystem in Fig. 5.1 kann nun aus der Steigung ein ΔE
zugeordnet werden. Es gilt genähert $\Delta E \simeq 2.6\ m_0 c^2$. Dieser Wert ist
übrigens auch mit der mittleren Energie des Positronenspktrums konsistent. Innerhalb der relativ großen Fehler stimmt die Skalierung (5.5)
gut mit dem Experiment überein. Das ist bemerkenswert, denn nach theoretischen Rechnungen (Rei O 78) spielen sich dem Einstufenprozeß kohä-
rent überlagernde Zweistufenprozesse eine Rolle, die allerdings im
wesentlichen um 90° in der Phase gedreht sind (vgl. auch Kan 78). Weiterhin ist keinesfalls als gesichert anzusehen, daß ΔE für ein Stoßsystem als Funktion von $2\hat{t}$ konstant bleibt.

Eines der wichtigsten Ergebnisse dieser Untersuchung ist, daß die
überkritischen Systeme von dieser Skalierung innerhalb der experimentellen Fehler nicht abweichen. Zur gleichen Schlußfolgerung gelangt man
beim Vergleich des Positronenspektrums vom überkritischen U+U mit dem

unterkritischen U+Pb System (siehe Fig. 5.2), die sich statistisch signifikant voneinander nicht unterscheiden. Für diesen experimentellen Befund gibt es verschiedene Erklärungsmöglichkeiten. Es könnte sein, daß das $1s_\sigma$-Niveau nicht oder nicht tief genug in das negative Energie-kontinuum eintaucht, um einen meßbaren Effekt zu machen. Weiterhin wäre denkbar, daß trotz Eintauchens des $1s_\sigma$-Niveaus keine signifikante Ab-weichung vom exponentiellen Verlauf von P_e+ als Funktion der Stoßzeit $2\hat{t}$ bzw. in den Positronenspektren auftritt in Übereinstimmung mit kürz-lich durchgeführten Rechnungen (Rei M 80).

Die theoretischen Rechnungen für P_e+ sind in Fig. 5.1 einge-zeichnet. Sie beschreiben bei kleinen Stoßzeiten $2\hat{t} \simeq 1.7 \cdot 10^{-21}$ s das Experiment gut, bei größeren Zeiten $2\hat{t} \simeq 2.7 \cdot 10^{-21}$ s scheint es ge-ringfügige Abweichungen in den U+Pb und U+U Systemen zu geben. Die spektrale Verteilung der Positronen stimmt für das U+Pb Stoßsystem gut mit den theoretischen Rechnungen überein (vgl. Fig. 5.2), während für das U+U System zwischen 0.5 MeV und 1.3 MeV Abweichungen auftreten, die aber noch einer weiteren experimentellen Bestätigung bedürfen. In einem kürzlich wiederholten Experiment wurde gefunden, daß die Rechnung im gesamten Energiebereich um ca. einen Faktor 1.6 über der Messung liegt, die spektrale Form aber gut wiedergibt.

Für $2\hat{t} = 1.75 \cdot 10^{-21}$ s ist in Fig. 5.1 im Zusatzbild die Z_u-Ab-hängigkeit von P_e+ dargestellt. Sie kann durch

$$P_e+ \propto Z_u^{\sim 20.3} \qquad (5.6)$$

beschrieben werden. Wiederum ist aus einer derartigen Darstellung keine Signatur für ein Eintauchen des $1s_\sigma$-Niveaus in das negative Energie-kontinuum beim U+U und U+Cm Stoßsystem zu erkennen.

Zusammenfassend kann gesagt werden, daß es schwierig ist, aus den energieintegrierten Positronenerzeugungswahrscheinlichkeiten P_e+ rein experimentell unter Verwendung von Extrapolationen von unterkritischen Systemen her Aussagen über das Eintauchen oder Nichteintauchen des $1s_\sigma$-Niveaus zu erhalten. Aussichtsreicher scheint es zu sein, die spek-trale Verteilung eines unterkritischen Systems mit derjenigen eines überkritischen Systems zu vergleichen. Dazu muß die Meßgenauigkeit aber erheblich verbessert werden.

6.   Perspektiven der Positronenspektroskopie

Die Experimente zur Positronenspektroskopie sind aus zweierlei Gründen mit ca. 30 % noch relativ ungenau. Einmal sind die Erzeugungswahrscheinlichkeiten pro gestreutem Teilchen sehr klein ($< 10^{-3}$) , so daß es bereits problematisch ist, in der zugeteilten Strahlzeit genügend Statistik zu sammeln. Andererseits macht die genaue Berücksichtigung des nuklearen Positronenanteiles wegen der Unkenntnis der Multipolarität Schwierigkeiten. Das Problem der Statistik kann durch verbesserte Apparaturen mit größerer Nachweiswahrscheinlichkeit gelöst werden. Für die in Kapitel 3.2 beschriebene Halbleiterzähleranordnung zur Messung von Positronenspektren müßte sich die Nachweiswahrscheinlichkeit bei Verwendung rechteckiger Zähler mit den Abmessungen 35 x 23 mm , die sternförmig angeordnet werden könnten, um einen Faktor 2 bis auf 13 % verbessern lassen. Es wird gegenwärtig versucht, derartige Zähler zu bauen. Ein weiteres Solenoid-Transportsystem bei der GSI (sogenanntes variables Solenoid (Bal B 80)) hat bei Verwendung einer Spiralblende zur Elektronenunterdrückung (Bac B 75) bereits eine Nachweiswahrscheinlichkeit von 12 %. In diesem Zusammenhang sind auch Verbesserungen am Positronenzähler des Orangenspektrometers (Ber B 80) zu erwähnen. Mit einem neuen Zähler ist die Koinzidenz mit der 511 keV Vernichtungsstrahlung nicht mehr erforderlich, wodurch bei einer relativen Impulsakzeptanz $\Delta p/p \simeq 15$ % , der volle relative Raumwinkel $\Delta\Omega/\Omega = 0.2$ erreicht wird. Schließlich kann mit einem im Bau befindlichen sogenannten Torispektrometer bei vollständiger Abtrennung des Elektronenuntergrundes für die Positronen eine Nachweiswahrscheinlichkeit von 20 % erwartet werden. Hierbei handelt es sich um einen S-förmig gekrümmten Solenoiden, in dessen toroidalem 1/r-Magnetfeld Elektronen und Positronen in entgegengesetzte Richtungen driften. Am Ende einen 1/4-Torus können die Elektronen dann ausgeblendet werden und die Positronen driften im anschließenden 1/4-Torus auf die ursprüngliche Position zurück, so daß sich dieses Gerät bezüglich seiner Abbildungseigenschaften für Positronen wieder annähernd wie ein Solenoid verhält.

Die drei zuletzt genannten Experimente verwenden ortsauflösende Parallelplattenzähler zum Nachweis der gestreuten Teilchen in kinematischer Koinzidenz. Bei unsymmetrischen Stoßsystemen (z. B. U+Pb) kann dann der Stoßparameter eindeutig bestimmt werden.

Mit der Erhöhung der Nachweiswahrscheinlichkeit für Positronen scheint eine Verbesserung der Statistik in den Positronenspektren um einen Faktor 5-10 möglich, so daß bei der Messung von Positronenspektren eine Genauigkeit von besser als 10 % in einem 100 keV breiten Energieintervall erreicht werden sollte. Die Genauigkeit der Messung wird dann wahrscheinlich durch die Korrektur auf nukleare Untergrundpositronen begrenzt. Beträgt der Anteil an nuklearen Positronen 40 % und fordert man, daß auf Grund dieser Korrektur das atomare Positronenspektrum auf 5 % genau sein soll, so muß er auf 7.5 % genau bestimmt werden, was sicher nicht einfach ist. Ob diese Genauigkeit auf dem eingeschlagenen Weg erreicht werden kann oder ob ein channeling Experiment (Kau K 78) den Ausweg liefert, muß die Zukunft zeigen.

Alle bisher beschriebenen Experimente wurden bei Energien unterhalb der Coulombbarriere durchgeführt, d. h. die stoßenden Kerne sollten nahezu reine Rutherford Trajektorien durchlaufen (was allerdings bei 5.9 MeV/u nicht mehr sicher ist). Derartige Experimente haben den Nachteil, daß die Aufenthaltsdauer des $1s_\sigma$-Niveaus im negativen Energiekontinuum sehr kurz ist (ca. 10^{-21} s).

Wie von Rafelski et al. (Raf M 78) vorgeschlagen wurde, sollte bei Zeitverzögerung nach einem tiefinelastischen Stoß von einigen 10^{-21} s die spontane Positronenerzeugung zu einer scharfen Struktur bei der Energie des eintauchenden Niveaus führen. Neuere Rechnungen zeigen ausgeprägte Oszillationen in den Positronenspektren ähnlich wie sie für δ-Elektronenspektren bei Zeitverzögerung vorhergesagt wurden (Sof R 79).

Erste experimentelle Schritte in diese Richtung wurden in diesen Experimenten unternommen. Es wurde die Positronenerzeugung in Koinzidenz zu Spaltfragmenten gemessen, die in Rückwärtsrichtung von Halbleiterzählern registriert werden (siehe Fig. 3.5). Durch diese Nachweistechnik werden sehr kleine Stoßparameter selektiert und es wird auf Grund der Kinematik ausgeschlossen, daß es sich um Spaltung infolge Reaktion mit leichten Targetverunreinigungen (O oder C) handelt. Die totale Positronenerzeugungswahrscheinlichkeit (nicht korrigiert auf atomaren Untergrund) ist in Fig. 6.1 dargestellt. Die e^+-Erzeugungsrate für U+Pb und U+U bei 5.9 MeV/u ist konsistent mit reiner nuklarer Positronenerzeugung wie ein Vergleich mit der ^{248}Cm Spaltquelle zeigt. Daraus kann die Schlußfolgerung gezogen werden, daß bei dieser Energie nur das Projektil bzw. ein Kern spaltet. Bei 7.5 MeV/u nimmt die Positronenerzeugung für das U+U und U+Cm System um einen Faktor 3 zu. Daraus kann aber nicht gefolgert werden, daß ein atomarer (vielleicht spontaner) Anteil an Positronen beobachtet wurde, weil die zugehörigen γ-Spektren ebenfalls um denselben Faktor ansteigen (vgl. Fig. 6.2). Der mit Null verträgliche Effekt an atomarer Positronenerzeugung kann mit destruktiven Interferenzen der Positron-Amplituden im ein- und auslaufenden Kanal (Rei S 79, Kan 79, Rei G 80) zusammenhängen, wenn die Zeitverzögerung infolge Kernkontakts nur in der Größenordnung von $1 \cdot 10^{-21}$ s liegt (Wol 77, Sch T 78).

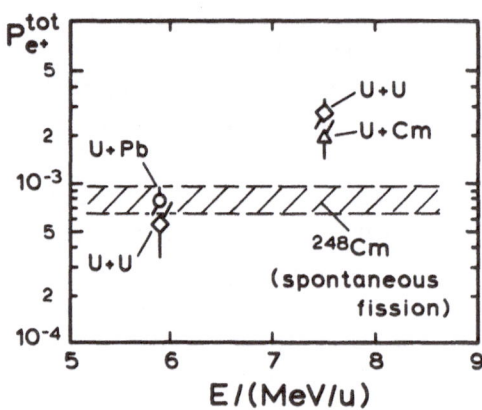

Fig. 6.1

Gesamtzahl der Positronen pro in Rückwärtsrichtung nachgewiesenem Spaltfragment $P_{e^+}^{tot}$ (vgl. die experimentell Anordnung in Fig. 3.5) beim Beschuß von ^{208}Pb und ^{238}U mit ^{238}U Projektilen einer Energie von 5.9 MeV/u und 7.5 MeV/u . Für den ^{248}Cm Punkt wurden die Spaltfragmente in Vorwärtsrichtung mit dem Plastik-Zähler (siehe Fig. 3.5) nachgewiesen, wobei der Nachweis von quasielastisch gestreuten Teilchen durch den Energieverlust in einer 4 mg/cm dicken Berylliumfolie unterdrückt wurde. Die Größe $P_{e^+}^{tot}$ wurde nicht auf nuklearen Untergrund korrigiert. Der schraffierte Bereich stellt eine Messung von $P_{e^+}^{tot}$ für eine ^{248}Cm Spaltquelle dar (Bac B 81).

Die hier beschriebenen neueren Experimente wurden in Zusammenarbeit mit W. Bonin, W. Engelhardt, E. Kankeleit, M. Mutterer, P. Senger, F. Weik, R. Willwater sowie V. Metag und J.B. Wilhelmy durchgeführt, denen ich zu großem Dank verpflichtet bin. Insbesondere möchte ich auch J. Foh danken, der wesentliche Teile der Impulselektronik entwickelte, ohne die diese Experimente undenkbar gewesen wären.

Diese Arbeit wurde mit Mitteln des Bundesministeriums für Forschung und Technologie und der GSI Darmstadt unterstützt.

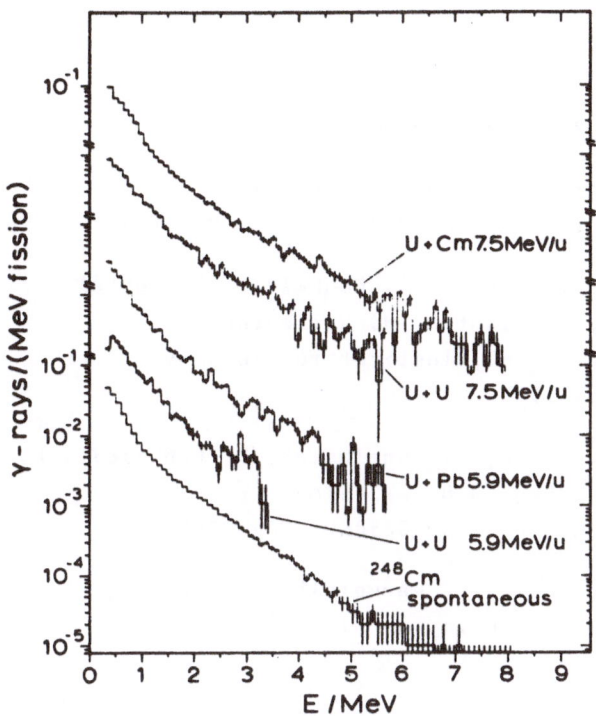

Fig. 6.2

Die zu Fig. 6.1 gehörigen γ-Spektren, die mit einem 7.5 x 7.5 cm NaJ-
Detektor aufgenommen wurden (vgl. Fig. 3.5).

Literaturverzeichnis

Arm K 79   P. Armbruster, P. Kienle
             Z. Physik A 291, 399 (1979)

Bac 78   H. Backe, in: Trends in Physics 1978,
             ed. M.M. Woolfson, Adam Hilger Ltd., Bristol 1979, p 445

Bac B 75   H. Backe, H. Bokemeyer, E. Kankeleit, E. Kuphal,
             Y. Nakayama, L. Richter, R. Willwater
             Laborbericht Nr. 67, Institut für Kernphysik der
             TH Darmstadt, 1975

Bac B 80   H. Backe, W. Bonin, W. Engelhardt, E. Kankeleit, P. Senger,
             F. Weik, V. Metag, J.B. Wilhelmy
             to be published and GSI Scient. Rep. 80-3, p. 101

Bac B 81   H. Backe, W. Bonin, E. Kankeleit, L. Richter, F. Weik,
             V. Metag, J.B: Wilhelmy et al.
             to be published

Bac H 78   H. Backe, L. Handschug, F. Hessberger, E. Kankeleit,
             L. Richter, F. Weik, R. Willwater, H. Bokemeyer, P. Vincent,
             Y.Nakayama, J.S. Greenberg
             Phys. Rev. Letters 40, 1443 (1978)

Bal B 80   A. Balanda, H.J. Beeskow, K. Bethge, H. Bokemeyer,
             H. Folger, J.S. Greenberg, H. Grein, A. Gruppe, S. Ito,
             S. Matsuki, R. Schulé, R. Schultz, D. Schwalm, J. Schweppe,
             R. Steiner, P. Vincent, M. Waldschmidt
             GSI Scient. Rep. 80-3, p. 161

Ber B 80   E. Berdermann, F. Bosch, M. Clemente, F. Güttner, P. Kienle,
             W. Koenig, C. Kozhuharov, B. Martin, W. Potzel, B. Povh,
             Ch. Tsertos, W. Wagner, Th. Walcher
             GSI Scient. Rep. 80-3, p. 103

Bok 79   H. Bokemeyer
             Atomic Physics with Heavy Ions
             GSI-Report 79-4 and Heavy Ion Physics.
             Eds.: A. Berind, V. Ceausescu, I.A. Dorobantu,
             Proceed. Predeal International School 1978, p. 489

Heß 77   F.P. Heßberger
             Untersuchungen zur Nachweiswahrscheinlichkeit eines
             Solenoid-Beta-Spektrometers für Positronen
             Diplomarbeit 1977, Institut für Kernphysik der TH Darmstadt

Kan 78   E. Kankeleit, in: Nuclear Interactions, ed. by B.A. Robson,
             Lecture Notes in Physics. Vol. 92.
             Springer-Verlag, New York 1979, p. 306

Kan 79   E. Kankeleit, in: Proceedings of the Twelfth Summer School
             of Nuclear Physics, Mikołajki,
             Poland, 1979, Nukleonika 25, 253 (1980)
             and Institut für Kernphysik, Technische Hochschule
             Darmstadt, Report 79/20

Kau K 78   K.G. Kaun, S.A. Karamyan
             JINR P 7-11420, Dubna 1978 and GSI-tr-9/78
             (Übersetzung P. Strehl)

Koz K 79   C. Kozhuharov, P. Kienle, E. Berdermann, H. Bokemeyer,
             J.S. Greenberg, Y. Nakayama, P. Vincent, H. Backe,
             L. Handschug, E. Kankeleit
             Phys. Rev. Letters 42, 376 (1979)

Lie A 80   D. Liesen, P. Armbruster, F. Bosch, S. Hagmann,
             P.H. Mokler, H.J. Wollersheim, H. Schmidt-Böcking,
             R. Schuch, J.B. Wilhelmy
             Phys. Rev. Letters 44, 983 (1980)

Mig 77   A.B. Migdal:
             Qualitative Methods in Quantum Theory.
             (London and Amsterdam: Benjamin; Ontario, Sydney and Tokyo:
             Don Mills), 1977, p. 115

Raf F 78   J. Rafelski, L. Fulcher, A. Klein
             Phys. Rep. 38 C, 227 (1978)

Raf M 78   F. Rafelski, B. Müller, W. Greiner
             Z. Physik A 285, 49 (1978)

Rei B 80   J. Reinhardt, W.Betz, P. Gärtner, J. Kirsch, U. Müller,
             T. de Reus, K.H. Wietschorke, G. Soff, B. Müller,
             W. Greiner
             Proceedings of the XVIII International Winter Meeting on
             Nuclear Physics, Bormio 1980, p. 816

Rei G 77   J. Reinhardt, W. Greiner
           Rep. Progr. Phys. 40, 219 (1977)

Rei G 80   J. Reinhardt, W. Greiner, B. Müller, G. Soff:
           Electronic and Atomic Collisions,
           Eds.: N. Oda, K. Takayanagi, Morth-Holland Publishing
           Company, Amsterdam 1980, p. 369

Rei O 78   J. Reinhardt, V. Oberacher, B. Müller, W. Greiner, G. Soff
           GSI-Bericht-M-8-78

Rei S 79   J: Reinhardt, G. Soff, B. Müller, W. Greiner:
           Erice School on Heavy Ion Interactions at High Energies,
           Erice (Italy), 26 March - 6 April, 1979

Sch S 78   P. Schlüter, G. Soff, W. Greiner
           Z. Physik A 286, 149 (1978)

Sch T 78   R. Schmidt, V.D. Toneev, G. Wolschin
           Nucl. Phys. A 311, 247 (1978)

Sof R 78   G. Soff, J. Reinhardt, W. Betz, J. Rafelski
           Phys. Scripta 17, 417 (1978)

Sof R 79   G. Soff, J. Reinhardt, B. Müller, W. Greiner
           Phys. Rev. Letters 43, 1981 (1979)

Wei 80   F. Weik
           Dissertation 1980, Institut für Kernphysik der Technischen
           Hochschule Darmstadt

Wie M 79   K.-H. Wietschorke, B. Müller, W. Greiner, G. Soff
           J. Phys. B 12, L 31 (1979)

Wol 77   G. Wolschin
           Nukleonika 22, 1165 (1977)

M.D. Scadron

Advanced Quantum Theory and Its Applications Through Feynman Diagrams

1979. 78 figures, 1 table. XIV, 386 pages
(Text and Monographs in Physics)
ISBN 3-540-09045-2

This is a concise, yet comprehensive text on the basic techniques used in theoretical elementary particle physics. The central topic is relativistic Feynman diagrams and their construction in lowest order, applied to electromagnetic, strong, weak, and gravitational interactions.

A thorough discussion of relativistic Feynman diagrams is preceded by a review of transformation theory, used to formulate advanced quantum theory in group theoretical language. Next, scattering theory and its many applications to nuclear, atomic, and solid-state physics are presented. Finally, the finite parts of higher order graphs are calculated and the calculations justified in terms of renormalized field theory and dispersion theory.

The work provides a natural introduction to the use of relativistic Feynman diagrams, and increases the student's understanding of the natural forces by reducing them to their simplest and most general terms.

Springer-Verlag
Berlin
Heidelberg
New York

Lecture Notes in Physics

Selected Issues from
Lecture Notes in Mathematics